DK

Earth's Incredible Habitats
Mountain

Written by
Jason Bittel

Illustrated by
Sandra Neuditschko

Written by Jason Bittel
Illustrated by Sandra Neuditschko

Senior editor Olivia Stanford
Senior art editor Charlotte Bull
Additional editorial Becca Arlington,
Abi Luscombe, Rea Pikula
Designers Hannah Moore, Lucy Sims
Senior picture researcher Sakshi Saluja
Jacket designer Bettina Myklebust Støvne
Jacket illustrator Joe Stansbury
Jacket coordinator Elin Woosnam
Managing editor James Mitchem
Managing art editor Diane Peyton Jones
Senior production editor Nikoleta Parasaki
Production controller John Casey
Art director Mabel Chan
Managing director Sarah Larter

Biology consultant Dr Nick Crumpton
Geology consultant Dr Tom Argles

First published in Great Britain in 2024 by
Dorling Kindersley Limited
DK, One Embassy Gardens, 8 Viaduct Gardens,
London, SW11 7BW

The authorised representative in the EEA is
Dorling Kindersley Verlag GmbH. Arnulfstr. 124,
80636 Munich, Germany

Copyright © 2024 Dorling Kindersley Limited
A Penguin Random House Company
10 9 8 7 6 5 4 3 2 1
001–340591–Sept/2024

A CIP catalogue record for this book
is available from the British Library.
ISBN: 978-0-2416-6980-8

Printed and bound in China

www.dk.com

MIX
Paper | Supporting
responsible forestry
FSC™ C018179

This book was made with Forest
Stewardship Council™ certified
paper – one small step in DK's
commitment to a sustainable future.
Learn more at www.dk.com/uk/
information/sustainability

Contents

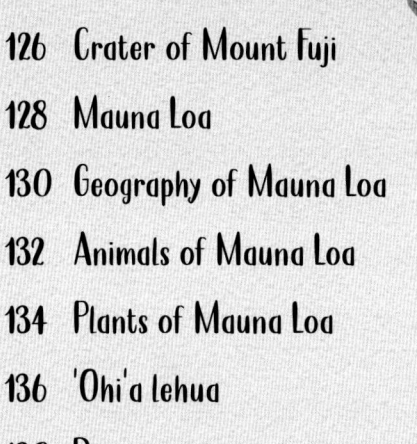

North America

The hazy Appalachian Mountains seen here are found in the east of North America, but they are not alone. In the west, the largest range on the continent, the Rocky Mountains, dominate. Further south, in Mexico, is the Sierra Madre mountain system. In the northwest, the Alaska Range towers. It holds the continent's highest point, Mount Denali, which has an elevation of 6,190 m (20,310 ft).

South America

The Andes are perhaps the most familiar mountain range in South America. They give rise to the mighty Amazon River. But did you know they contain the highest point in the Americas? That's Argentina's Mount Aconcagua. South America also holds stunning flat-topped mountains in the north, including Mount Roraima, and the awe-inspiring highlands of the Brazilian Plateau, seen here.

Europe

Europe is known for the snow-crested Alps, but it also boasts the Scandinavian Mountains in the north, the Pyrenees between Spain and France, the Carpathians in Central Europe, and the Caucasus Mountains – which are found at the continent's border with Asia. These ranges are home to many animals, such as these griffon vultures flocking in the Pyrenees.

Africa

You're looking at a sunset on Africa's tallest mountain – Mount Kilimanjaro in Tanzania. As cooling air creates cloud below its peak, the mountain looks like an island at sea. This is just one of the continent's wondrous high places, which also include the Ethiopian Highlands in the east, the Atlas Mountains in the north, and the Drakensberg Mountains in the south.

Asia

There are many mountain ranges to be found in Asia – the world's largest continent. The Ghats, seen here, are actually two mountain ranges that form a V-shape along India's coast. Other large ranges include the Urals in Russia, the Zagros Mountains in Iran, and of course, in the east, the Himalayas, which are home to more than 100 peaks of 7,000 m (23,000 ft) or more!

Oceania

Oceania is a region that includes an incredibly diverse array of islands in the Pacific Ocean, from Australia all the way to Hawaii. Perhaps surprisingly, even islands can have some seriously impressive mountains, including Australia's Great Dividing Range, Hawaii's mighty Mauna Loa volcano, and New Zealand's Southern Alps – seen here with a common mountain resident, the kea.

What is a mountain?

There is no worldwide definition of a mountain, though many refer to steep, rocky landforms that rise abruptly and impressively above their surroundings. A mountain in some countries would only be viewed as a small hill in others! While most people usually picture steep-sided mountains like the one in the picture below, this is just one kind of mountain you'll find in this book.

Mountain habitats

What kinds of plants and animals live on mountains? Well, that depends on what elevation you look at. Mountains can have many different kinds of habitat depending on how high they are and what kind of climate they have. The higher you go up a mountain, the fewer trees there are and the colder it becomes.

Alpine tundra

Alpine meadows

Shrubs and bushes

Coniferous forest

Deciduous forest

The elevation above which snow settles on the peak of a mountain is called the snow line.

As you move up a mountain, the habitats change in a similar way to if you moved towards the poles of the planet.

Trees can't grow above a certain elevation as it becomes too cold. This is called the tree line.

Mountain height

There are a few different ways you can measure the height of a mountain. The most common way is to measure the height from a mountain's peak to sea level. However, scientists can also measure how far a peak is from the seafloor.

8,848.86 m
(29,031 ft)

Sea level

Qomolangma Feng (Mount Everest)

Qomolangma Feng (Mount Everest) is the tallest mountain in the world if you measure from sea level to the highest point.

10,210 m
(33,500 ft)

Sea level

Seafloor

Mauna Kea

If you measure from its base at the bottom of the sea to its tallest point, then Mauna Kea is the highest mountain in the world.

Glaciers and rivers

When snow builds up year after year in mountainous regions, it can become squished down and hardened into ice, creating a glacier. Despite being made of ice, glaciers can move – albeit very slowly – thanks to gravity acting upon their huge mass. As they move down a mountain, they eventually melt, creating lakes and rivers.

Mountain species

Many plants and animals are found in mountain habitats and nowhere else. They have adapted to the cold, high elevation, and other challenges, and many are no longer able to survive in the world below.

Snow leopards are secretive animals that live among the peaks of Asia. They have warm fur to survive the cold.

You won't see edelweiss growing in a swamp, desert, or forest. This plant only grows on mountains.

Types of mountain

All mountains are tall, but they aren't all pointy. The shape of a mountain depends on where and how it formed. It is also affected by how long ago it was created, as mountains can change shape over time. Mountains are found everywhere, even at the bottom of the ocean, but they are often found together in ranges – we'll see why this is below!

Mountain formations

Mountains which are formed in different ways have different shapes. The effects of erosion over time can also change the shape of a mountain. Let's look at the various formations found in this book.

Mountain ranges

When many mountains form together, they are known as a range. They often form along plate boundaries in long, thin strips called mountain chains.

Plateaus

Plateaus are large areas of uplifted land that are flat on top, but they rise dramatically away from the world around them on at least one side.

Volcanoes

As lava spills out of gaps in the Earth's surface, it can create gigantic mountains, or volcanoes, seemingly out of nowhere! The type of lava affects their shape.

Tabletop mountains

Much like plateaus, tabletop mountains are flat on top, but they have a smaller area and stand alone. They're left behind as rock is eroded away from around them.

Plate tectonics

The Earth's surface is made up of massive, flat chunks of rock called tectonic plates. They fit together, but each can slowly move, so over time they can bump together or pull apart. The huge forces at these plate boundaries cause earthquakes and volcanic eruptions, and also push up mountains.

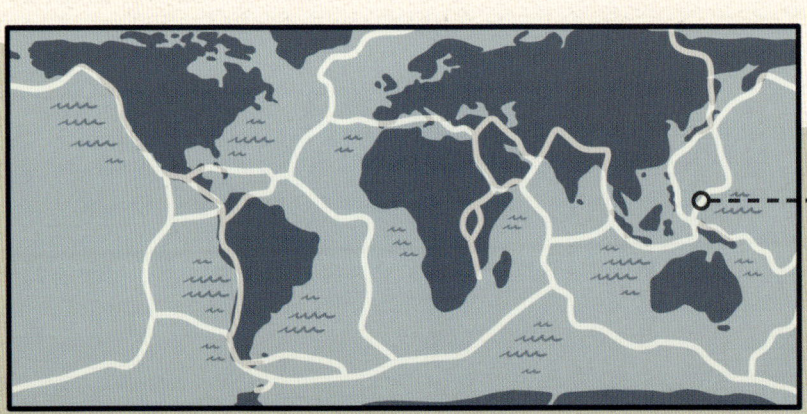

Plate boundaries

Plate boundaries

Most mountain building happens at the edges, or boundaries, of tectonic plates, as well as lots of other interesting geology. Plates can move in different directions, which affects what happens at the Earth's surface.

Divergent boundaries are where two plates move away from each other. Lava can rise up in the gap to form volcanoes.

Convergent boundaries occur where two plates collide, making Earth's crust thicker and forming mountain ranges.

When one plate slides beneath another, it is known as subduction. Volcanoes are common here, but mountains may be forced upwards too.

When plates slide past each other, it's known as a transform plate boundary. Earthquakes are common at this type of boundary.

Erosion

Given enough time, even the tallest mountains will crumble and shrink, thanks to the powerful effects of erosion. This process involves the weathering (wearing away) of rocks into smaller pieces by wind, running water, freezing, and even the grinding force of glaciers.

Rockies

Sierra Nevada
This mountain range is almost entirely in the U.S. state of California.

Mauna Loa

Sierra Madre
Three mountain ranges make up this mountain system in Mexico.

Appalachians
The Appalachian Mountains are some of the oldest mountains on the planet.

Pyrenees
The Pyrenees form a high, natural wall between Spain and France.

Atlas
These mountains are named after a titan, or giant, from Greek mythology.

Mount Roraima

Andes

Mountains of the world

If you were asked to name five mountains or mountain ranges, could you do it? This map reveals that the Earth is absolutely full of magnificent mountains, from the southern tip of the Andes in South America, all the way up to the Scandinavian Mountains in Norway, and many places in between. Which ones haven't you heard of?

Scandinavian
These mountains run right up to the sea, where they create beautiful bays called fjords.

Urals
The Urals are super-old, having formed about 275 million years ago!

Caucasus
Historically, these mountains were used as a boundary between Europe and Asia.

Alps

Tian Shan
Tian Shan means "the Celestial Mountains". They reach 7,439 m (24,406 ft) high.

Altai
The Altai Mountains contain the source of the Ob, a large river which flows north to the Arctic Ocean.

Mount Fuji

Qing Zang Gao Yuan

Himalayas

Zagros
These mountains are home to striped hyenas, leopards, and brown bears.

Ghats
Running along India's coast, these mountains are divided into the Eastern Ghats and the Western Ghats.

Ethiopian Highlands

Drakensberg
These mountains contain human rock and cave art that is several thousand years old.

Great Dividing Range

Types of rock

If you go to a museum, you will learn there are tons of different kinds of rock. However, every single one of them can be put into one of just three categories.

Igneous

Igneous rocks form when magma or lava cools – which can happen either on the Earth's surface, making extrusive igenous rocks, or deep underground, making intrusive igneous rocks.

Sedimentary

When specks of rock, shell, or grains of sand collect together, for instance on the seabed, they can be compacted to form sedimentary rocks. You can often see layers in these rocks.

Metamorphic

Like a caterpillar turning into a butterfly, these rocks form when another kind of rock changes because of extreme pressure or heat under the ground.

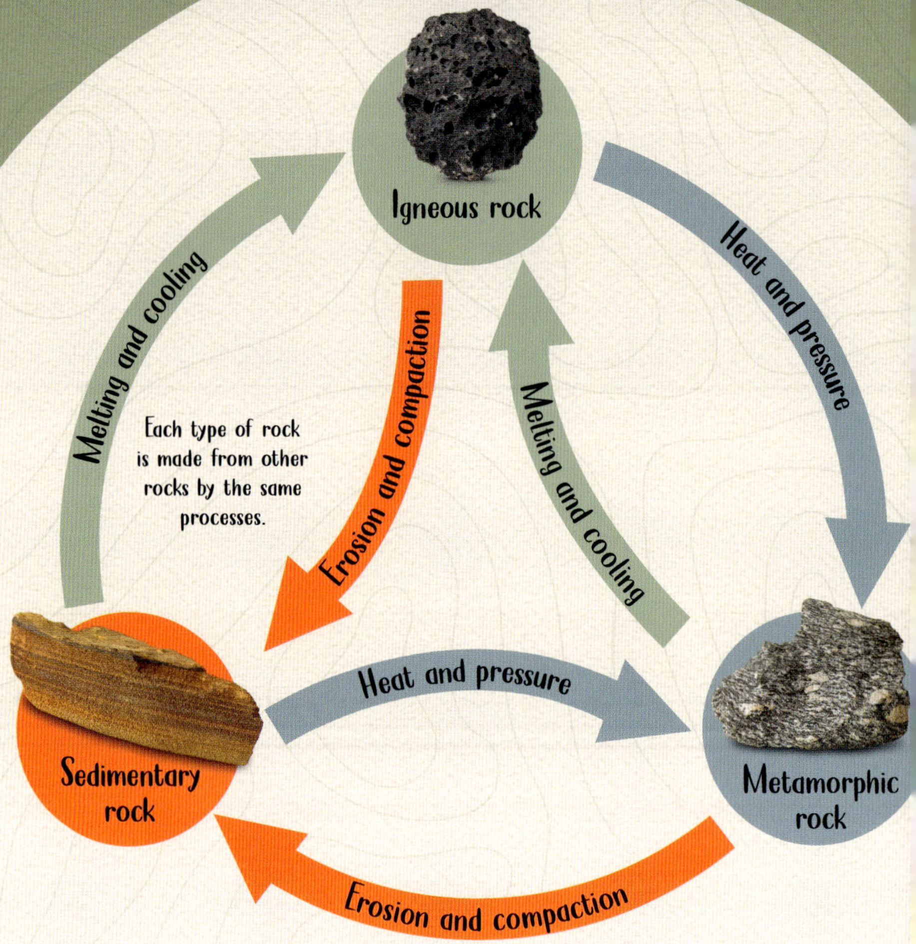

Igneous rock

Melting and cooling

Heat and pressure

Erosion and compaction

Melting and cooling

Each type of rock is made from other rocks by the same processes.

Sedimentary rock

Heat and pressure

Metamorphic rock

Erosion and compaction

The rock cycle

Even though they can be ancient, each kind of rock can be changed into the other kinds through processes such as heating, pressure, melting, cooling, and erosion. Scientists call these ongoing processes the rock cycle.

Rocks

Mountains are made from rocks, but the kinds of rocks and the way they are arranged can reveal all sorts of interesting things about how and when a mountain was formed. From volcanoes made from cooled lava to ocean sediments pushed up to the tops of peaks during mountain building, every stone and pebble you come across has a story – some of which go back billions of years.

Inside the Earth

No one has ever been to the Earth's centre, but experiments tell us that our planet has four layers. The metal core has a solid inner part and a liquid outer layer. The mantle and crust are both solid rock, but the lower mantle flows very slowly. The rigid upper mantle and crust are divided into tectonic plates.

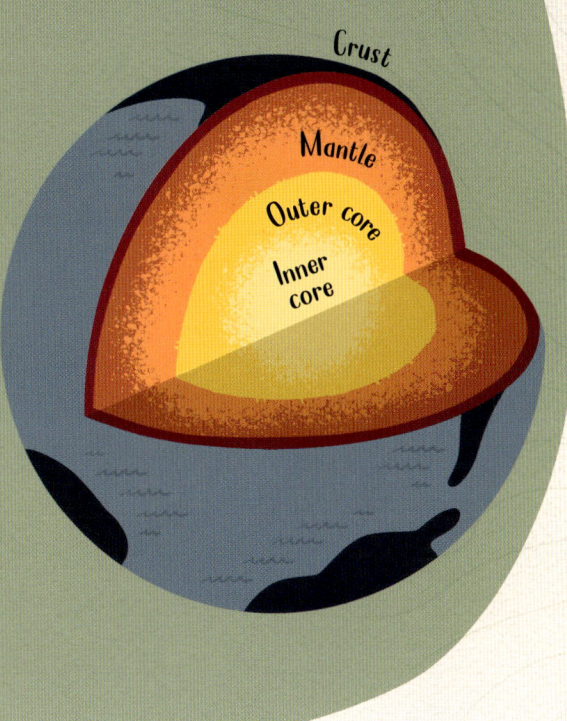

Crust
Mantle
Outer core
Inner core

Fossils

When most organisms die, their tissues decay and disappear. But every so often, these remains get covered in sediment and become fossilized. Over millions of years, minerals seep into the remains and replace the tissues. This means fossils are not bones, shells, or feathers any more, they are rocks!

Magma and lava

While both terms refer to molten, or melted, rock, there's one key difference that separates magma from lava – location! Scientists call molten rock that is underground magma, while lava is used to describe molten rock that breaks through to the Earth's surface at volcanoes.

New mountains can form from igneous rocks where lava cools.

Global warming

The Earth is getting warmer because of a process called global warming. When greenhouse gases are released into Earth's atmosphere, they trap heat from the Sun. While the planet's climate has always changed, it's now happening faster because of people burning fossil fuels and the cutting down of forests, which increase greenhouse gases.

Trapped heat

Atmosphere

Melting glaciers

Many glaciers today are left over from the last Ice Age about 10,000 years ago. But now that the planet is heating up, these giant mounds of ice are melting at an alarming rate. Take a look at the satellite images below showing the change in the Tsanfleuron Glacier in Switzerland.

2001

2022

Threats

Mountains are home to more than 85 per cent of the world's species of amphibians, birds, and mammals, with many living on mountains and nowhere else. Unfortunately, mountain habitats are under threat from a number of different issues, including climate change and pollution.

No longer adapted

Every spring, mountain hares shed their white fur coats for brown ones that help them to hide from predators on the mountainside. However, global warming is making snow melt earlier than the hares can change coats. This is just one way climate change is shifting the routines that animals have evolved over millions of years.

Higher tree lines

As the climate gets warmer, many mountain species have no choice but to move higher up the slopes to find a cooler area. Scientists have documented this shift with trees, watching as certain species claim higher and higher territories. The worry? This process can only go on so long before there is nowhere else to climb.

Microplastics

You might think that the top of a mountain is about as remote and unspoiled a location as you can get. However, scientists have shown that even the upper reaches of Qomolangma Feng (Mount Everest) are now affected by microplastic pollution – microscopic pieces of plastic can be found in the snow there.

Avalanches

When an unstable mass of snow and rock breaks loose and tumbles down a mountainside, it's known as an avalanche. While scientists aren't sure yet how a changing climate will affect avalanches worldwide, they have shown that changes to rain and snow can make avalanches worse in certain areas.

Mountain ranges

Mountains are landforms that rise above the land around them – and a mountain range is a series of such landforms that are close together. Mountain ranges can extend over vast distances. For instance, the Andes stretch across the entire length of South America, and are considered the longest continuous mountain range on Earth! We'll be exploring the Andes as well as the Himalayas, Alps, Rockies, and the Great Dividing Range.

FACT FILE

Length
8,900 km (5,500 miles)

Age
25 million years

Highest point
Mount Aconcagua
6,961 m (22,838 ft)

The Andes

The Andes are the world's longest mountain range on land, with a total length of around 8,900 km (5,500 miles).

When most people think about South America, they imagine the Amazon Rainforest or Amazon River. However, the Andes Mountains contain some of the most impressive landscapes in the world. Forming the continent's spine, this mountain range reaches all the way from Venezuela in the north down to the tip of Chile in the south. The Andes are made up of many different kinds of mountain habitats, from towering cliffs and snow-capped peaks to cool deserts and cloud forests.

High altitude

The Andes contain the largest mountains outside of Asia, with an average height of around 4,000 m (13,000 ft). The tallest peak, Mount Aconcagua, is found in Argentina.

Andean condors are one of the highest-flying birds on the planet. They can soar up to 6,500 m (21,300 ft)

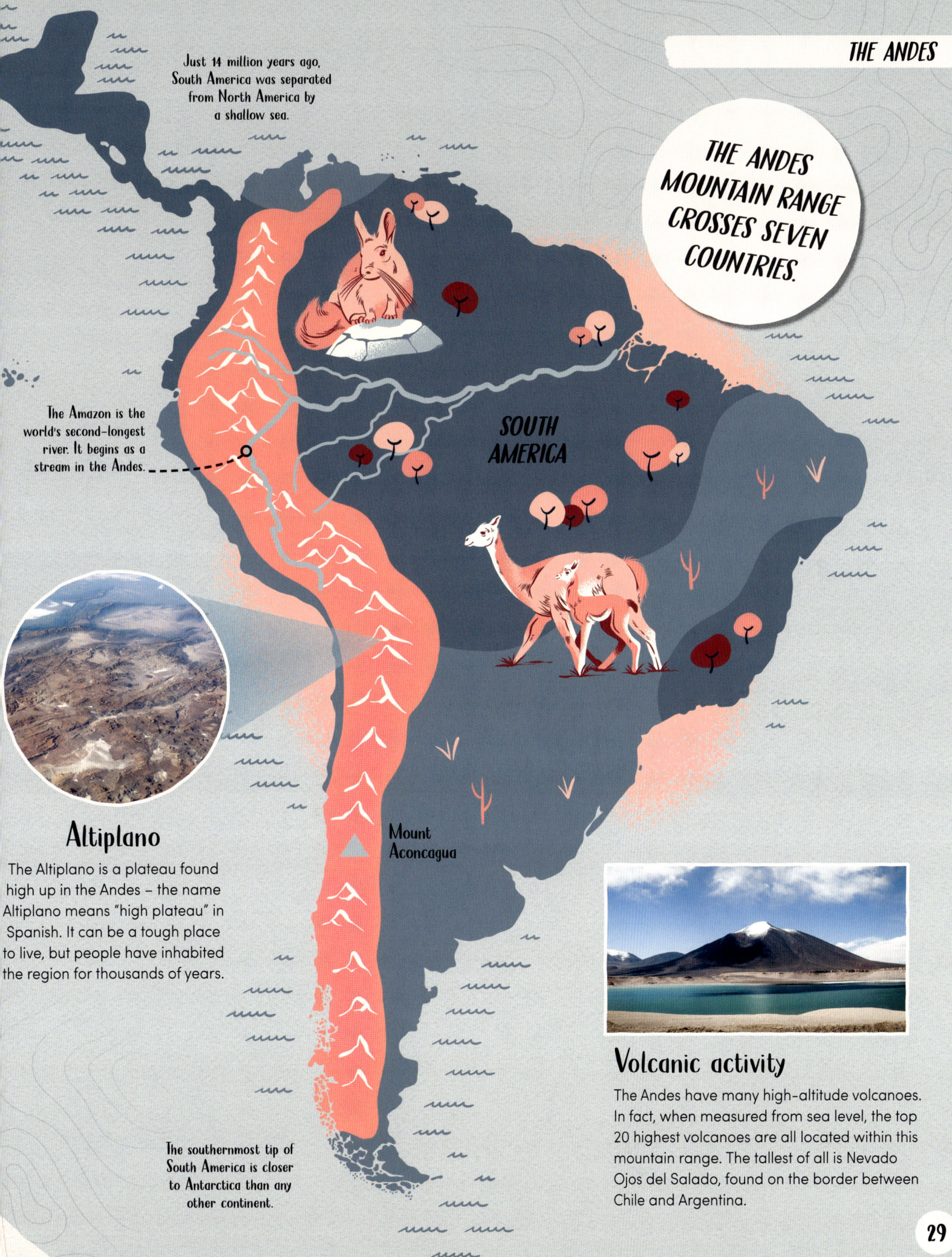

THE ANDES MOUNTAIN RANGE CROSSES SEVEN COUNTRIES.

Just 14 million years ago, South America was separated from North America by a shallow sea.

The Amazon is the world's second-longest river. It begins as a stream in the Andes.

SOUTH AMERICA

Altiplano

The Altiplano is a plateau found high up in the Andes – the name Altiplano means "high plateau" in Spanish. It can be a tough place to live, but people have inhabited the region for thousands of years.

Mount Aconcagua

The southernmost tip of South America is closer to Antarctica than any other continent.

Volcanic activity

The Andes have many high-altitude volcanoes. In fact, when measured from sea level, the top 20 highest volcanoes are all located within this mountain range. The tallest of all is Nevado Ojos del Salado, found on the border between Chile and Argentina.

Geography of the Andes

Over the course of millions of years, the Andes have risen up into the sky and now stand as the second-tallest mountain chain on Earth, after the Himalayas. And they're still getting bigger! Each year, due to the shifting of tectonic plates - gigantic pieces of the Earth's crust and mantle - the Andes grow a little over 1 cm (½ in). Magma from where these tectonic plates meet rises to feed many active volcanoes in this mountain range.

Where the two plates meet under the Pacific Ocean, a massive underwater trench has formed.

When one tectonic plate dives beneath another, it can push up mountains over the course of millions of years.

The three peaks known as Torres del Paine - the Blue Towers - are made of hard granite that originally formed underground from solidifying magma.

South American Plate

Nazca Plate

Known as the Nazca Plate, this plate is mostly under the Pacific Ocean and is moving east at a rate of 5 cm (2 in) each year.

The South American Plate includes most of the continent that shares its name, as well as a huge chunk of the Atlantic Ocean.

Formation

The Andes began to form when the Nazca Plate under the Pacific Ocean started to slide under the South American Plate. The colossal forces slowly wrinkle the rock and make it rise, like the rumples in a scrunched-up rug.

Salt flats

Found high up in the Andes, Bolivia's Salar de Uyuni is the largest salt flat on Earth. The area used to be covered in lakes, but a changing climate made most of the water evaporate, leaving behind a thick layer of salt and sediment.

Scientists use this extremely flat and highly reflective surface to test satellite cameras.

Mountain building

Friction caused by the collision of the two tectonic plates under the Andes creates earthquakes and also pathways for magma to rise up out of the Earth's mantle. Where that magma cools or reaches the Earth's surface as lava, new mountains can form.

Andesite is a dark rock that often has paler crystals within it.

Rock types

Much of the Andes is made up of igneous rocks – rocks formed from cooled lava and magma. One type that can be found all over the world but which gets its name from the Andes is andesite.

Animals of the Andes

Because the Andes stretch across an entire continent, these mountains include a wide range of habitats that support an amazing number of animal species. In fact, scientists have found around 600 mammal species, 1,700 bird species, and 400 species of fish. Even more impressively, the Andes are home to more species of amphibian than anywhere else on Earth.

Andean condor
(Vultur gryphus)

At up to 15 kg (33 lb), this bald bird is one of the heaviest flying creatures on the planet. With keen eyesight, and a 3 m (10 ft) wingspan, condors soar high in the sky in search of animals to eat – alive or dead.

Guanaco
(Lama guanicoe)

Like its cousins the llama, alpaca, and vicuña, the guanaco comes from the same family as camels. Thick wool helps guanacos survive the cold temperatures at high elevations, but has also made the animals prized targets for hunters.

Northern viscacha
(Lagidium peruanum)

Looking like a cross between a bunny and a squirrel, viscachas are a kind of rodent native to Peru. They have thick fur and live in burrows between rocks. Some colonies house up to 80 animals.

Spectacled bear
(Tremarctos ornatus)

Named for the glasses-shaped bands of colour around their eyes, spectacled bears are the only bears found in South America. Usually nocturnal and herbivorous, these bears are considered shy and avoid people whenever possible.

Southern huemel
(Hippocamelus bisulcus)

The huemel is an endangered member of the deer family with short, branching antlers. Most remaining southern huemels live in Chile, with small populations also surviving in Argentina.

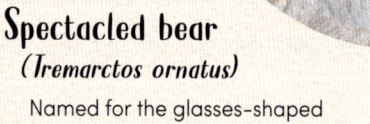

Extreme environments

The Andes are very different depending on where you go. For instance, in northwestern Colombia, you can stand in one of the wettest tropical rainforests in the world, where it can rain 300 days out of the year. Meanwhile, in Chile's Atacama Desert, more than a decade can pass without a single drop of rain, making it the driest place in the world!

Not many animals can live in the salty lakes found in the Atacama Desert, but many flamingos spend their summer there.

Flamingos of the Andes

There are six species of flamingo in the world, and three of them can be found in the Andes. Sometimes called the volcano flamingos, these birds have specialized bills that allow them to filter algae, brine shrimp, and other food out of the water from salty lakes.

James's flamingo

Chilean flamingo

Andean flamingo

Ichu grass
(*Stipa ichu*)

Sometimes called Peruvian feather grass, tufted ichu grass is a favourite food of llamas and other grazers, as well as a handy source of roofing material for local peoples. This grass is also used in gardens around the world.

Andean daisy
(*Rockhausenia nubigena*)

Andean daisies look like flowers you may find in the park. However, this species has a special reason for keeping its blooms low to the ground. This helps the Andean daisy protect itself from strong winds and low nighttime temperatures.

Plants of the Andes

Even though potatoes and tomatoes have become popular worldwide, both crops actually came from the Andes! These are just two of the more than 30,000 plant species that call this region home. Unbelievably, one out of every six plant species on Earth can be found in the Andes. Many plants here are adapted to bright sunlight and cold temperatures.

Queñoa de altura
(*Polylepis tarapacana*)

Queñoa de altura makes its home at the top of the world – no other tree species grows higher! However, growing at high elevations comes at a cost. The higher the tree grows, the shorter and more shrub-like it is.

Peruvian Andes hairy cactus
(*Austrocylindropuntia floccosus*)

Also known as polar bear cactus or wool cactus, this plant has a fluffy appearance that some say looks like a resting sheep. However, these hairs have a purpose, protecting the plant from bitter cold and harsh sunlight.

Potatoes

Quinoa

Mountain crops

While potato varieties are a critical part of Andean agriculture, many other food plants grow well in the direct sun and high altitude of these mountains. Maize or corn, quinoa, maca, amaranth, and kaniwa are all important crops in this region.

Potato plant
(*Solanum tuberosum*)

If you love chips or mashed potato, then you should be glad that around 10,000 years ago, people living in the Andes figured out that this poisonous plant had tasty, edible tubers attached to its roots! There are now many different varieties of potato.

Yareta
(*Azorella compacta*)

It looks like a bright-green blob, but yareta is actually a kind of cushion plant made up of many very tiny, very densely-packed leaves. Sometimes spelled llareta, these plants can live an amazing 3,000 years or more!

35

Queen of the Andes

The plant known as the Queen of the Andes is famous for growing to enormous sizes – up to 12 m (40 ft). It only flowers once, and only after reaching an age of 80 to 100 years! After sending up its huge flower spike, the impressive plant produces seeds, and then dies. Then the cycle repeats again.

Endangered

Like many of the plants that live in the harsh environments of the Andes, the Queen of the Andes is threatened by climate change as well as the fires people use to clear the land for crops and livestock. There are only a few areas where this plant now grows and it is endangered.

Pollination

It takes a lot of help to pollinate a tower of flowers this big, and no fewer than six different species of hummingbird have been seen buzzing around Queen of the Andes blooms. The plant also serves as food and shelter for many other animals.

Spiky leaves

For most of its life, all that can be seen of this plant is a short stem and a spray of pointed leaves. Curving spikes on the edges of its leaves may keep leaf-munching animals from bothering it.

The Red Lagoon

High up on Bolivia's Altiplano, there's a lake so red,
it doesn't seem like it could possibly be real. But it is!
Its vibrant crimson colour is the result of red algae and
other tiny microorganisms that live in it. This abundant
food attracts different species of flamingo. One of them,
known as the James's flamingo, is so rare that scientists
worry it might become extinct.

Length
2,500 km (1,550 miles)

Age
50 million years

Highest point
Qomolangma Feng
(Mount Everest)
8,848.86 m (29,031 ft)

The Himalayas

The Himalaya Mountains get their name from two Sanskrit words that mean "abode of snow".

Located in Asia, the Himalayas stretch across the countries of India, Pakistan, Afghanistan, China, Bhutan, and Nepal. The range can be divided into three zones, each with their own landscapes, rock types, and wildlife: the Greater Himalaya, Lesser Himalaya and Outer Himalaya. While these gigantic mountains often look lifeless, an array of resilient creatures and plants have found ways to call the "Roof of the World" their home.

The bar-headed goose migrates north across the Himalayas to its breeding grounds.

Tall peaks

Of the 14 mountain peaks that are over 8,000 m (26,200 ft), eight lie at least partly in landlocked Nepal, including the tallest mountain of all – Qomolangma Feng (Mount Everest).

Qomolangma Feng (Mount Everest)

The tallest mountain on the planet, Qomolangma Feng (Mount Everest) sits on the Nepal-China border. It's so high that oxygen levels at the summit are only one-third of those at sea level.

Qomolangma Feng (Mount Everest)

Snow and ice

Remember how Himalaya means "abode of snow"? Well, the Himalaya Mountains contain huge amounts of snow and ice, including many glaciers.

Active glaciers

Glaciers are huge masses of ice that form when snow falls at the tops of mountains and gets squished into ice over many centuries. They flow down valleys in the high mountains like rivers in slow motion. The Himalayas are home to more than 15,000 glaciers.

Found in Nepal, the Ngozumpa Glacier is the longest glacier in the Himalayas.

Geography of the Himalayas

Older mountains are worn down by erosion, through wind, rain, snow, and ice, but relatively "new" mountains like the Himalayas appear sharper and more jagged. High peaks in the Himalayas act like a barrier to clouds being blown northwards, which release their water as they meet the mountains. This deluge of rain feeds 19 rivers that flow mainly to the south of the Himalayas, while to the north, there is desert.

The sea in the sky

Millions of years ago, the rocks that now make up the Himalayan peaks used to lie beneath an ocean. That's why today you can find sedimentary rocks, such as limestone, filled with fossils of fish and other sea creatures at the top of the mountains!

Ammonites were ancient sea creatures with spiral shells. This ammonite fossil was found high in the Himalayas.

Limestone is a sedimetary rock that forms on the bottom of oceans. As the Himalayas were forced up, layers of limestone were lifted upwards.

Today

38 million years ago

55 million years ago

71 million years ago

History of the Himalayas

About 70 million years ago, the land that now forms India and its surroundings was an island in the middle of a vast ocean. However, movement of the tectonic plate beneath it slowly carried the landmass north until it crashed into the coast of Asia, helping to form the Himalayas.

Formation

The Himalayas are some of the youngest mountains, forming throughout the last 50 million years. These mountains are still rising up into the sky where the Indian and Eurasian Plates meet, and rock is being crumpled and forced upwards.

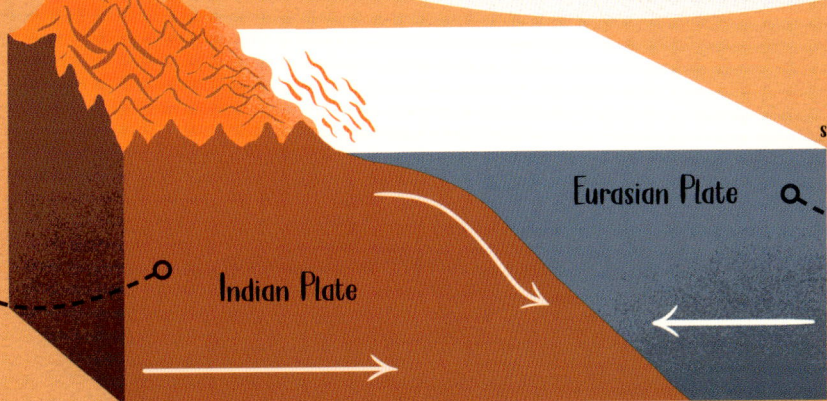

The Indian Plate is being slowly pushed downwards, or subducted, underneath the Eurasian Plate.

Eurasian Plate

Indian Plate

The Indian Plate is being crumpled and forced upwards, forming the Himalaya Mountains above.

Bearded vulture
(Gypaetus barbatus)

While lots of vulture species have bald heads, this one has a beard! Well, that's what they call the bird's head feathers, at least. To break large animal bones, bearded vultures drop them out of the sky and then swallow the pieces.

Himalayan marmot
(Marmota himalayana)

This large burrowing rodent is considered an ecosystem engineer because of the way its underground tunnels churn up the soil and allow plants to take root. Himalayan marmots are also a favourite food for brown bears and snow leopards.

Animals of the Himalayas

From mountain peaks and ice fields to warm valleys and raging rivers, the Himalayas hold a variety of ecosystems for animals to call home. Life is tough, but plenty of species have evolved to survive in a place full of extremes. A few popular survival strategies include warm winter coats, strong wings for migration, and sturdy hooves for climbing.

Himalayan jumping spider
(Euophrys omnisuperstes)

This tiny jumping spider lives at a higher altitude than any other spider on Earth. Its scientific name means "highest of all" – how fitting! Experts believe Himalayan jumping spiders survive by hunting bugs blown up high by the wind.

Himalayan tahr
(Hemitragus jemlahicus)

This relative of goats and sheep can weigh up to 140 kg (310 lbs). Like mountain goats, the Himalayan tahr's hooves have a rubbery insole that allows them to cling to the mountainside.

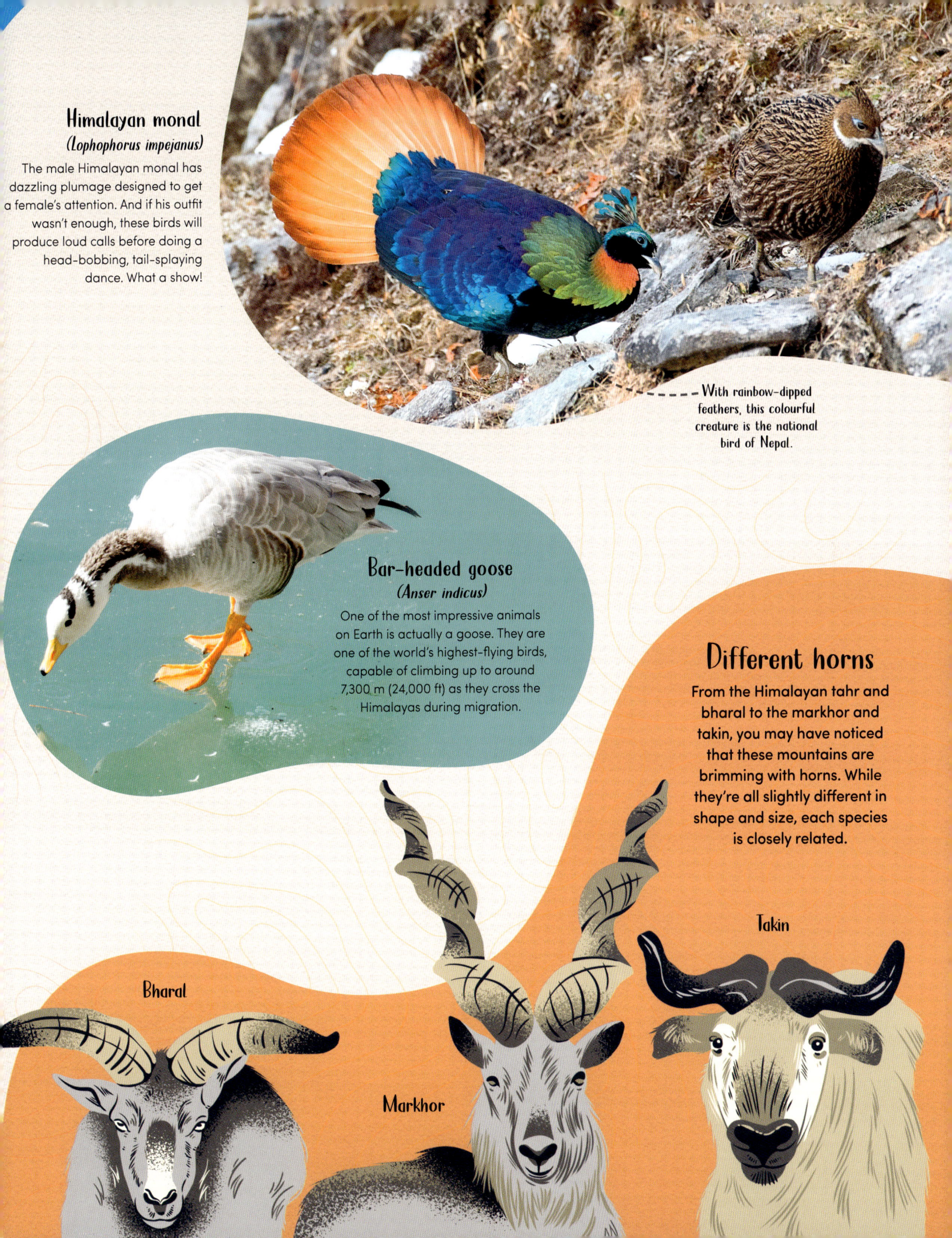

Himalayan monal
(Lophophorus impejanus)

The male Himalayan monal has dazzling plumage designed to get a female's attention. And if his outfit wasn't enough, these birds will produce loud calls before doing a head-bobbing, tail-splaying dance. What a show!

With rainbow-dipped feathers, this colourful creature is the national bird of Nepal.

Bar-headed goose
(Anser indicus)

One of the most impressive animals on Earth is actually a goose. They are one of the world's highest-flying birds, capable of climbing up to around 7,300 m (24,000 ft) as they cross the Himalayas during migration.

Different horns

From the Himalayan tahr and bharal to the markhor and takin, you may have noticed that these mountains are brimming with horns. While they're all slightly different in shape and size, each species is closely related.

Takin

Bharal

Markhor

Snow leopard

The snow leopard is one of the most elusive and difficult to study animals on the planet. These mammals are usually solitary, and inhabit remote and dangerous terrain, stalking and pouncing on everything from marmots to mountain sheep. Interestingly, despite the big cat's name, scientists believe the snow leopard is more closely related to tigers than other leopards.

Spots

You might think an animal that lives in the snow would want to be all white, but snow leopards hunt in both snowy and rocky areas. Their greyish-black spots make them harder to see.

Master of disguise

Snow leopards are so difficult to spot in the distance and so wary of people that scientists have better luck finding the cats' footprints and poo than the animals themselves.

Footprints

When you live on top of the world, snowshoes come in handy. Extra large, heavily-furred paws help snow leopards walk on top of the snow rather than sinking into it. They leave large footprints in the snow.

The tree line

On really tall mountains, you can see that there are fewer trees and other vegetation the higher up you go. Eventually, there are no trees at all. This is known as the tree line, which refers to the point at which there is no longer enough oxygen, warmth, or water for trees to grow.

Plants of the Himalayas

While we tend to think of a mountain range as just one thing – big, snowy cliffs – this isn't the case. The Himalayas are actually tropical at the bottom of the mountains, then subtropical, temperate, and alpine as you move upwards. More than 4,000 species of tree call these mighty mountains home.

Bumblebees have furry bodies to stay warm in the cold.

Snowball plant
(*Saussurea gossypiphora*)

This is what a snowball would look like if it could blossom. Found at high altitudes, local people use the wool of the snowball plant to seal wounds and stop cuts from bleeding.

Himalayan blue poppy
(*Meconopsis grandis*)

Imagine going hiking through the mountains and stumbling upon a field of bright blue flowers. Known as the Himalayan blue poppy, these delicate blooms like shady meadows, slopes, and woodlands.

Sandwort
(*Eremogone bryophylla*)

While this little bundle of white blooms doesn't look like much, sandwort grows at altitudes of up to 6,180 m (20,280 ft). It is considered to be the highest-growing flower on Earth.

Drumstick primula
(*Primula denticulata*)

Also known as lollipop primulas because of their shape, these plants seem to disappear in the winter months. But they are actually just hiding, and their flowers bloom beautifully in the spring.

Noble rhubarb
(*Rheum nobile*)

Noble rhubarb stands nearly 2 m (7 ft) tall and is topped with ghostly white leaves. Rather than making energy from the sun, like green leaves do, these ones keep the plant warm and protect it from ultraviolet radiation!

Pollination at altitude

Way up in the mountains, there are fewer birds and other large animals to pollinate flowers. So some plants have developed a trick – they make smaller flowers with sweeter nectar to entice tiny pollinators like flies and bumblebees.

Lenticular clouds

Did you know that mountains can actually affect the shape of clouds? Lenticular clouds, also known as lee wave clouds, form downwind of a huge obstacle, such as a mountain. As the wind forces air up the mountain, it cools and any water vapour in it condenses into a cloud. Lenticular clouds are disc-shaped and seem to hover in the same place, making them look like flying saucers!

Length
1,200 km (750 miles)

Age
100 million years

Highest point
Mont Blanc
4,807 m (15,771 ft)

The Alps

The Alps form a crescent shape that stretches for approximately 1,200 km (750 miles) across southern and central Europe.

With snow-capped peaks and exceptionally beautiful views, the Alps are one of the tallest and most prominent mountain ranges in Europe. They run across parts of Austria, France, Germany, Italy, Liechtenstein, Monaco, Slovenia, and Switzerland. Humans have been living in this region for at least 50,000 years, and many mythical creatures have been said to lurk on the alpine summits, too. Of course, the real-life creatures found in the Alps, from alpine ibex to golden eagles, are just as glorious!

Jura Mountains

North of the Alps, there's another, younger mountain range called the Jura Mountains. These mountains gave the dinosaur-packed Jurassic Period its name. The limestone that makes up the Jura Mountains was laid down on the seabed during that time and contains many fossils.

Iconic species

The Alps are home to a vast number of species. In Europe, only the Mediterranean region has more plants and animals to its name. Some are found nowhere else in the world.

Edelweiss is a plant adapted to the high altitudes of the Alps.

Mont Blanc

Glacial lakes

When glaciers melt and that water gets trapped at high elevations, it can form massive glacial lakes. For example, Lake Garda was formed by glacial melt. It's the largest lake in Italy!

Geography of the Alps

If you could look down at the Alps from up in space, you'd see a crescent-moon shaped mountain range covered in snow, with green valleys cutting into its side. We know this because scientists can show us what the Alps look like from above using satellites! While some mountain ranges are still rising and others are shrinking, the Alps appear to be staying the same height.

Nappes

When two tectonic plates are forced together, stresses build up in the rocks until their layers buckle and bend. This forms folded patterns in huge slabs of rock, called nappes. The Alps are famous for these wave-like formations.

Valleys

The Alps are full of valleys carved out by icy glaciers. For instance, the largest glacier in the Alps, the Aletsch Glacier, covers a whopping 82 km² (32 miles²) and has created a huge valley as it has moved downhill. There are also extraordinary scratch marks where the glacier dragged stones across the valley floor.

Formation

The Alps started to form about 65 million years ago, right after Tyrannosaurus rex lived out its last days. The African Plate collided with the Eurasian Plate and caused mountains to emerge.

The African Plate moved northwards and was pushed underneath the Eurasian Plate.

Eurasian Plate

African Plate

The colliding plates drove up rocks that had been buried beneath an ancient sea to create massive mountains.

The fold of a nappe can be clearly seen in the Dent de Morcles mountain in Switzerland.

The rock dolomite is made from a mineral also called dolomite.

Gneiss contains many folds from being put under huge stress.

Rock types

Two common rocks of the Alps are dolomite and gneiss. Dolomite is a sedimentary rock that gives its name to a part of the Alps called the Dolomites. Gneiss is a metamorphic rock made when granite and other rocks are buried and heated intensely during mountain building.

Golden eagle
(*Aquila chrysaetos*)

With a wingspan of up to 2 m (7 ft), these massive birds soar above the Alps in search of rabbits, marmots, or other small mammals to eat. Golden eagles usually nest on cliffs, another reason why they feel so at home in the Alps.

Alpine ibex
(*Capra ibex*)

Also known as a steinbock or bouquetin, this wild goat has huge, curving horns used for displays and battles between males. Thanks to special hooves and powerful legs, ibex are able to cross vertical cliffs.

Animals of the Alps

With birds and bats in the skies, bears and wolves on the slopes, and fish and salamanders in the lakes and glacial streams, scientists have discovered around 30,000 animal species living in the Alps. Like the alpine ibex, many are endemic, which means they are found in one place and nowhere else!

Alpine chamois
(*Rupicapra rupicapra*)

Alpine chamois can run as fast as 50 kph (31 mph) across the rocky outcrops of the Alps. These animals are a type of goat-antelope and they can also leap 2 m (7 ft) straight up into the air, which helps them to escape from lynx, wolves, and bears.

Winter colours

Some animals change their colours throughout the seasons. Ptarmigans go from brown to white in the winter. This is to help them blend in with their snowy surroundings and hide from hungry predators.

Male in summer

Male in winter

Alpine newt
(Ichthyosaura alpestris)

These amazing amphibians have red bellies, paddle-like tails, and a striking, spotted appearance that makes them popular pets. In fact, a fondness for Alpine newts has led to the animals being introduced to new places, such as the United Kingdom.

Alpine marmot
(Marmota marmota)

To survive low temperatures at high altitudes, these rodents may hibernate for up to 200 days of the year. Alpine marmots like to snuggle up with their friends and family to stay warm, and up to 20 of them can be found hibernating together.

Apollo butterfly
(Parnassius apollo)

Get a good look at the Apollo butterfly while you can – these animals are one of the most endangered butterflies in Europe. The good news is scientists are working hard to save these insects, which are native to mountain meadows and pastures.

Plants of the Alps

Ibex and marmots are well-known alpine species, but these herbivores rely on grasses and plants for food. Fortunately, the Alps are home to more than 4,500 species of plants, each with their own adaptations for a life at altitude. For instance, one small plant known as net-leaved willow only grows 10–15 cm (4–6 in) high to avoid cold winds. But below ground, its roots can grow extremely long!

Alpenrose
(Rhododendron ferrugineum)

Despite its name, this mountain shrub is not a rose, but a rhododendron. It is also an evergreen, which means it does not lose its leaves, even in the coldest of winters. In late spring and early summer, alpenrose erupts in bright pink flowers.

Trumpet gentian
(Gentiana acaulis)

Growing low to the ground, the trumpet gentian sure knows how to stand out in a crowd. This species has brilliant blue flowers, each splaying outwards in a striking, flared tube that looks like the end of a trumpet.

Swiss pine
(*Pinus cembra*)

Some know the Swiss pine as the King of the Alps because these conifers can live up to 1,000 years old. This makes the Swiss pine one of the longest-living tree species on Earth.

Alpine snowbell
(*Soldanella alpina*)

Walk through the Alps in early spring, and you might just see alpine snowbells poking up through the melting snow. They are one of the first flowers to appear and their dark stems may even help to melt the snow around them.

Glacier buttercup
(*Ranunculus glacialis*)

At elevations of up to 4,270 m (14,000 ft), glacier buttercups grow higher than any other flowering plant – in the Alps, at least. These plants can also take root in scree fields with little soil or nutrients, earning them the nickname of "scree-creeper".

Summer meadows

If you're looking for wildflower hikes, the Alps offer some of the best on the planet. As spring gives way to summer, flowering plants bloom and make the mountain meadows explode with colour. The plants make the most of the warm summer sun by producing seeds before dissapearing for the winter.

Edelweiss

While edelweiss only ever grows about 30 cm (12 in) tall, this plant has become world-famous. It is an iconic plant of the Alps and the national flower of Austria and Switzerland. Today, you can find edelweiss growing in both the Alps and the Pyrenees, with other closely-related species found from northern Russia all the way to New Zealand.

Under protection

Due to its popularity over the last century, tourists have picked so many edelweiss flowers that experts are worried that the plant could actually become extinct. Climate change is also a threat to this species, so it is now protected in several countries.

Furry leaves

Edelweiss is sometimes called the "wool flower", thanks to its furry leaves. It lives in cold, windy, high-elevation areas, and its fluffy hairs protect its flowers from harsh sunlight and chilly weather.

Symbolic flower

Edelweiss has many different meanings. Some used to think the flower had magical powers, but mostly it has come to be a symbol of strength and bravery. Images of edelweiss are sometimes used in logos.

The Matterhorn

The Matterhorn is one of the best-known mountains on Earth, in part because it is shaped like a pyramid. Despite being found on the border between Switzerland and Italy, the rocks that formed the Matterhorn are thought to have originally been a part of the African tectonic plate. The word matterhorn means "peak in the meadow" in German. Because of its height, the top of the Matterhorn receives the first rays of sun at dawn before the valleys below.

Length
4,800 km (3,000 miles)

Age
75 million years

Highest point
Mount Elbert
4,401 m (14,440 ft)

The Rockies

Home to grizzly bears and mountain lions, the Rocky Mountains stand as one of the most impressive features in all of North America.

From Canada to the southern United States, the Rocky Mountains – also known as the Rockies – dominate the western side of the continent. As a relatively young mountain range, the Rockies have yet to be worn down too much by erosion, and therefore stand tall, sharp, and bold against the western sky. Mount Elbert, in the American state of Colorado, is the mountain range's highest point.

The mountain lion, also known as a puma or cougar, is a top predator.

Mountain predators

The predators found in the Rocky Mountains include mammals, such as mountain lions, magnificent birds, such as bald and golden eagles, and reptiles, like the western diamondback rattlesnake.

FROM A DISTANCE, THE ROCKIES CAN ACTUALLY LOOK PURPLE.

Mount Elbert

The Great Divide

The Great Divide is an imaginary line that runs along the Rockies and defines where precipitation will end up. Rain and snow that fall to the west of the line will flow into the Pacific Ocean, while to the east, they flow to the Atlantic Ocean.

Colorado Plateau

The Colorado Plateau is a large, high-altitude region bordered by the Rocky Mountains to the north and east, a lowland area called the Great Basin to the west, and the Sonoran Desert to the south.

Geography of the Rockies

You might say the Rocky Mountains started forming 3.7 billion years ago, when their oldest rocks were created. Or, when earlier mountain ranges formed in this region at 1.7 billion years ago and around 285 million years ago. But most people go with an age of 75 million years, because that's when the Rockies as we know them today were formed.

Ancestral Rocky Mountains

About 285 million years ago, before the Rockies as we know them existed, a much smaller mountain range rose up in the same area. The Ancestral Rocky Mountains then eroded away leaving behind sedimentary rocks in their place.

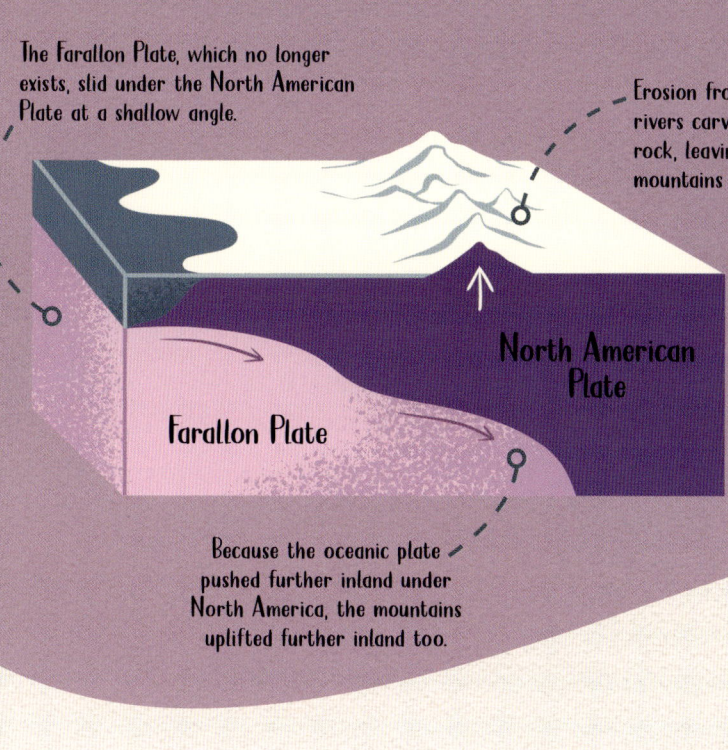

The Farallon Plate, which no longer exists, slid under the North American Plate at a shallow angle.

Erosion from glaciers and rivers carved away soft rock, leaving behind the mountains we see today.

North American Plate

Farallon Plate

Because the oceanic plate pushed further inland under North America, the mountains uplifted further inland too.

Formation

Unlike most mountain ranges, which form where tectonic plates meet, the Rockies are found further into the North American Plate. So how did they form? Scientists think the movement of an ancient tectonic plate, called the Farallon Plate, off the coast of California triggered other geologic action far inland.

Burgess Shale

Located in Canada's British Columbia, the Burgess Shale is found in an area of the Rockies made from ancient rocks. It is teeming with super-old fossils from the Cambrian Period, which lasted from 542 to 485 million years ago.

Anomalocaris was an ancient ocean predator that reached 60 cm (2 ft) long. It looked a little like a very large shrimp.

Fossils

The fossils found in the Burgess Shale offer a snapshot of what life looked like more than half a billion years ago, when sea life started to transform like never before! The fossil above shows one of the spiky mouthparts of a creature called Anomalocaris.

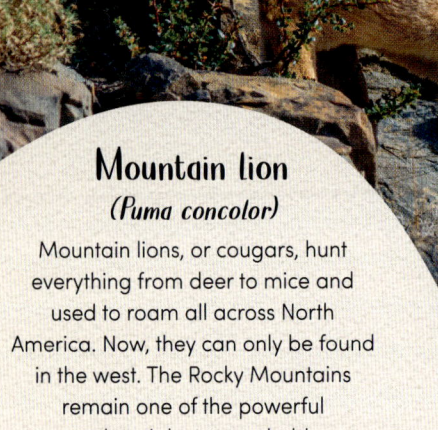

Mountain lion
(*Puma concolor*)

Mountain lions, or cougars, hunt everything from deer to mice and used to roam all across North America. Now, they can only be found in the west. The Rocky Mountains remain one of the powerful predator's last strongholds.

Greenback cutthroat trout
(*Oncorhyncus clarkii*)

Named for its green scales and a splash of red by its gills, this fish was thought to be extinct. Happily, it was rediscovered in the 1950s, but the greenback cutthroat trout is still in danger of disappearing.

Animals of the Rockies

It's a rough and tumble life in the Rockies, but the animals that have adapted to the harsh conditions are some of the most iconic creatures on Earth. There are bears, moose, beavers, rattlesnakes, and eagles, but also many smaller creatures that have come up with clever ways to survive high up among these tough peaks.

American pika
(*Ochotona princeps*)

These relatives of the rabbit are some of the few animals found above the tree line. Always thinking ahead, pikas pick plants and then dry them out in the sun before storing them below ground to be eaten throughout the winter.

Western toad
(*Anaxyrus boreas*)

Sometimes called the boreal toad, these amphibians produce toxins in their skin that help keep predators away. However, some foes, such as ravens, have learned they can flip the toads over to avoid their poison glands.

The western tanager migrates to the Rockies to breed in the summer.

Breeding

Wintering

Peregrine falcon
(Falco peregrinus)

Peregrine falcons may be small, but they can dive at speeds of up to 320 kph (200 mph) and kill other birds with a lightning-quick strike – all in midair! These aerial attacks are known as stoops.

Migration

Millions of birds, like this brightly-coloured western tanager, migrate north through the Rocky Mountains each year. Some will stay and spend the summer in the mountains, while others are just stopping through on their way up to Canada or even the Arctic.

Bighorn sheep
(Ovis canadensis)

When male bighorn sheep square off against one another – look out! These powerful mammals can bang their horns together at upwards of 32 kph (20 mph), creating a crashing sound that echoes through the mountains. The battles can go on for hours!

Kids

Baby mountain goats, known as kids, can walk, run, and climb within minutes of being born. Nannies and their kids form groups of up to 20 animals to help avoid predators, such as mountain lions and eagles.

Mountain goat

Despite their common name, scientists say these animals aren't actually true goats. They're more like goat-antelopes. But whatever you call them, these impressive mammals have found a way to survive way up on the mountaintops where few other animals go. Unlike lots of other horned animals, it's the females, called nannies, that are in charge of mountain goat society.

Salt lick

The grasses and other plants mountain goats eat are low in certain nutrients, such as salt, so these animals go to great lengths to find other naturally-occuring sources, like salt licks – rocky areas that contain minerals with salt in. Mountain goats will even fight bighorn sheep for the right to access salt licks!

Sure-footed

Strong and sturdy hooves with rubbery pads in the middle allow mountain goats to hold their grip on a variety of surfaces, including sheer rock faces. They have toes that spread wide for balance and powerful legs that help propel them up and down ledges, too.

Glacier lily
(Erythronium grandiflorum)

Pollinated by bumblebees and a prized food of grizzly bears and hummingbirds, glacier lilies are one of the first flowers to poke out as the snow melts. Interestingly, one other thing can make glacier lilies bloom – forest fires!

Plants of the Rockies

If you learned about one species of plant native to the Rocky Mountains every day, it would take you 13 years to get through them all. That's because there are more than 5,000 species! Fortunately, we've highlighted a few of the more colourful varieties you might spot there.

Mountain avens
(Dryas octopetala)

This white flower may look delicate but it is actually one of the toughest plants in the Rockies. Found in extreme places where only mountaineers tend to visit, a mountain avens can survive cold temperatures and rocky slopes. Its pollen has even been used to track glaciers!

Moss campion
(Silene acaulis)

Known as a cushion plant, moss campion makes soft, pillow-like mounds along the ground. The leaves of the mound protect its flowers from the cold and wind, until they all bloom at once. The purple flowers are tiny!

Rocky mountain iris
(Iris missouriensis)

Hummingbirds love rocky mountain irises, which are also sometimes called western blue flags. But beware the roots of this plant – they are poisonous and can make people sick when touched.

Harebell
(Campanula rotundifolia)

In the past, people thought the presence of harebell meant the presence of hares, or rabbits. But the truth is that harebell is just a very common plant and can be found at elevations of up to 3,600 m (11,800 ft).

Globe anenome
(Anemone multifida)

Globe anemone is a member of the buttercup family, and is a common wildflower throughout the Rockies. The plants can grow in rocky terrain and at altitudes of up to 4,260 m (14,000 ft).

North- and south- facing slopes

In the Rockies, slopes that face north receive quite a bit less sunshine than those that face south. They are also cooler and retain more water. As a result, north-facing slopes usually have more, taller trees that compete for sunlight, while south-facing slopes support smaller, wider trees with larger roots to find water.

North facing slope

South facing slope

Cloud valleys

Tall mountains like the Rockies can affect the weather of an area by changing the way air flows. For instance, when air passes over a mountain range, it cools. This is especially true at night. As the temperature drops, so does the air, descending down the mountainside and piling up in valleys. And if that air is moist, it can create thick, low-lying clouds or fog.

Length
3,600 km (2,237 miles)

Age
300 million years

Highest point
Mount Kosciuszko
2,228 m (7,310 ft)

The Great Dividing Range

The Great Dividing Range gets its name for the way it divides Australia's eastern coast from the rest of the island.

The Great Dividing Range stretches from Dauan Island in the north right down to the southeast coast of Australia. It's the fifth-longest mountain range on Earth. Interestingly, these mountains affect how weather moves. When winds from the east and rain clouds hit the mountain range, they drop most of their precipitation on the eastern side, creating freshwater for 11 million people, including those living in major Australian cities such as Melbourne, Canberra, Sydney and Brisbane.

Endangered animals, such as the pink robin and the mountain pygmy possum, can be found in the Great Dividing Range.

Rare animals

The Great Dividing Range is home to more than 70 per cent of Australia's threatened species. Unfortunately, climate change could cause many animals and plants found here to go extinct.

The Blue Mountains

The Blue Mountains are a part of the Great Dividing Range that are found near Sydney. These mountains get their colourful name from the blue tinge hanging over the area when viewed from a distance. It is thought that oil from eucalyptus tree leaves is what causes this blue haze.

Many species, such as the southern corroboree frog, are found in the Great Dividing Range and nowhere else on Earth.

Australian Alps

This smaller range is part of the southern end of the Great Dividing Range. When Europeans arrived in Australia, they thought the snow-capped peaks here looked a lot like the Alps in Europe, which is how these mountains got their name.

Mount Kosciuszko

Geography of the Great Dividing Range

Sometimes known as the Eastern Highlands, the Great Dividing Range is made up of plateaus and mountains. It runs from north to south, more or less following Australia's eastern coast. It is also the continent's largest mountain range. While gorgeous natural areas are a big draw for tourists, this area is also used for agriculture, lumber, mining, and energy projects.

Pacific Plate

The Pacific Plate reaches from Southeast Asia all the way to the western coast of North America.

The Australian Plate is much larger than the continent, stretching into the Indian and Pacific oceans.

Australian Plate

Formation

Unlike most mountain ranges, which are formed by collisions between two tectonic plates or volcanic activity, the Great Dividing Range seems to have been pushed up from below in two stages, due to changes in Earth's mantle below the Australian Plate.

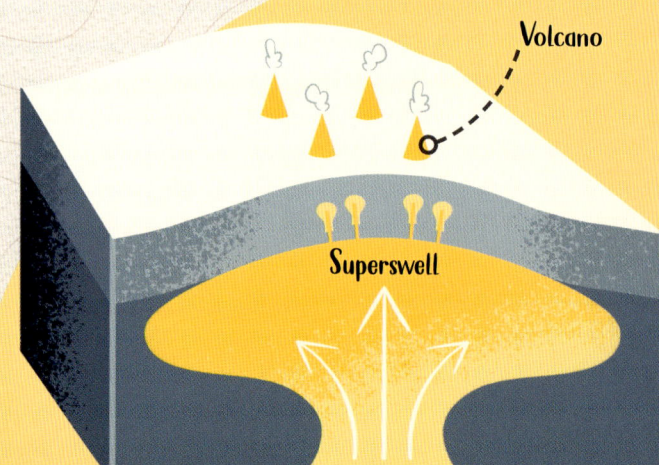

Volcano

Superswell

The Murray River is Australia's longest river, flowing for 2,530 km (1,570 miles).

River source

Many Australian rivers have their source in the Great Dividing Range, including the Snowy River on the eastern side. To the west, the Goulburn, Murrumbidgee, Lachlan, and Darling rivers meet up to form the huge Murray River. Around 120 million years ago this river flowed east, but the formation of the mountains has forced it to now flow west.

Granite forms underground when magma cools. It has large crystals of different minerals within it.

Gold is a shiny metallic element. It dissolves at depth in hot fluids that then rise and drop it nearer the Earth's surface.

The Pacific Superswell

It is thought that the Great Dividing Range was created by two events. The first was when the Australian Plate rebounded from being pushed under the Pacific Plate. The second was when the Autralian Plate drifted over the Pacific Superswell – an area of hotter mantle that pushes up the crust above.

Rocks and minerals

Large areas of the Great Dividing Range were formed from volcanic activity that started half a billion years ago. Both granite and gold are found in areas with volcanic activity. The gold here caused a series of gold rushes, where many people arrived to look for gold, during the continent's colonization.

79

Mountain pygmy possum
(Burramys parvus)

These small possums are Australia's only hibernating marsupial. Each spring, they wake up and search for bogong moths to eat. Unfortunately, their food supply is dwindling and this means that the mountain pygmy possums may disappear, too.

Animals of the Great Dividing Range

While lots of people probably think about kangaroos bounding through the desert when imagining Australia, these animals can be found in the mountains, too! As can wallabies, koalas, wombats, and a number of other well-known Australian icons. However, there are many other creatures you might not know that make their home here, even on Australia's tallest peak, Mount Kosciuszko.

Alpine water skink
(Eulamprus kosciuskoi)

Found in moss beds and wet heathlands in the mountains, these lizards dive into nearby water when danger is near. They also tunnel into the soft soil, which helps them survive the snow.

Bogong moth
(Agrotis infusa)

Every spring, bogong moths hatch in the lowlands and then wing their way up to 1,000 km (620 miles) to spend the summer in the cool of the caves found in the Great Dividing Range. They return to the lowlands to lay their eggs – if mountain pygmy possums don't gobble them up first!

Snow blanket

Mountain pygmy possums hibernate for up to seven months a year, and they do so under a thick blanket of snow! The snow layer can be up to 4 m (13 ft) thick. Interestingly, the snow actually insulates these marsupials from the harsh weather outside.

Mountain pygmy possums burrow into the snow, where the temperature is more constant.

The possums slow their heart rate and lower their body temperature when hibernating, and can lose half of their body weight!

Kosciuszko galaxias
(Galaxias supremus)

This fish is found in small streams around 1,900 m (6,230 ft) up Mount Kosciuszko. Like other mountain galaxias, it probably survives on a diet of molluscs, crustaceans, worms, and spiders. This species is critically endangered.

Pink robin
(Petroica rodinogaster)

This stunning little bird can only be found in southeastern Australia, where it lives in eucalyptus forests. It builds nests out on the end of moss-covered tree branches and migrates to lower altitudes in the winter.

Southern corroboree frog

About as big as a grape, the southern corroboree frog looks more like a plastic toy than a real, living creature, due to its glistening skin and black and yellow pattern. It lives high in the Great Dividing Range, around 1,760 m (5,770 ft) up. Don't get too close though – these tiny amphibians produce a toxin in their skin that makes them poisonous.

Bright colours

These frogs are the only amphibians to make their own poison rather than getting it from their food, and their bright colours are meant as a warning to predators! When animals are boldly coloured because they are dangerous, it is called aposematism.

Unusual eggs

Unlike lots of other frogs, which lay hundreds or thousands of eggs at a time, each corroboree frog female lays around just 16 to 38 eggs. Unusually, the eggs can be laid out of water in moist places, but when the rains arrive, their nest floods and the tadpoles hatch.

Forest fires

Unfortunately, only a few dozen corroboree frogs remain in the wild, and the species is endangered. Its two biggest threats are bushfires and an infectious fungus known as chytrid, which can be fatal to amphibians.

Plants of the Great Dividing Range

The plants of the Great Dividing Range include eucalyptus and acacia trees, but also a range of hummock grasses and dainty, beautiful flowering plants. One of the oldest and rarest trees on Earth, the Wollemi pine, is found here. Scientists actually thought it went extinct two million years ago, until it was rediscovered in 1994.

Mountain plum pine
(Podocarpus lawrencei)

One of Australia's few native conifer trees, the mountain plum pine produces tiny red fruits that mountain pygmy possums love. Interestingly, this plant can grow as a shrub or a full-height tree that reaches 15 m (49 ft) tall.

Snow gum
(Eucalyptus pauciflora)

There are more than 700 kinds of eucalyptus tree, but the snow gum is one of the toughest of them all. It grows in the mountains and can survive winter temperatures of –20°C (–4°F).

Anemone buttercup

Endangered plants

Like many unique places on Earth, the Great Dividing Range is home to species that exist nowhere else and are now in danger of disappearing. The anemone buttercup, silky daisy-bush, and bogong daisy-bush are all threatened by destruction of their habitat.

Silky daisy-bush

Bogong daisy-bush

Billy buttons
(Pycnosorus globosus)

What looks like one large, circular flower is actually a bunch of tiny clusters of yellow flowers, with downy white material underneath. Billy buttons, which are a member of the daisy family, grow in the Snowy Mountains, in the south of the Great Dividing Range during the summer.

Snow daisy
(Brachyscome nivalis)

Often found in snow gum forests, the snow daisy likes to grow in fields, damp places, and crevices between rocks. Snow daisies are perennials, which means the same plant will grow back year after year.

The Three Sisters

This incredible rock formation is found in the Blue Mountains near Sydney. Around 200 million years ago, these pillars, known as the Three Sisters, would have been buried in other rocks deep beneath an ocean. Over time, the region was pushed up and erosion slowly removed the softer rocks around them, leaving behind the towers we see today.

Plateaus

While mountains tend to have pointy peaks, a plateau is a landform that is both high in elevation and flat on top. Plateaus often rise sharply away from the lands below. When pieces of a plateau are separated because of erosion, they are known as outliers. Qing Zang Gao Yuan in Asia is considered the largest plateau on Earth, while the Ethiopian Highlands form the largest plateau in Africa.

Area
600,000 km²
(232,000 miles²)

Age
75 million years

Average height
3,200 m (10,500 ft)

Ethiopian Highlands

Home to almost half of Africa's tallest mountains, the Ethiopian Highlands are actually two mountain ranges separated by the East African Rift.

The Ethiopian Highlands are a stunning and important region in northeastern Africa that cover a large portion of Ethiopia. Sometimes called the Abyssinian Highlands, these mountains' slopes are home to millions of people and many wild animal species that cannot be found anywhere else. The region also contains more than 50 volcanoes, several of which are active. The land is raised into plateaus and peaks that contain high grassland and other mountain habitats.

High plateau

While none of the peaks in the Ethiopian Highlands are permanently snow-capped, this area is very high compared to the surrounding areas. It is Africa's largest high plateau.

Mountain species such as the giant lobelia thrive in the Ethiopian Highlands.

AFRICA

East African Rift

One of the most tectonically active regions in the world, the East African Rift is where three plates are moving away from each other. This has created a gap that separates the Western and Eastern Highlands – and they are still pulling apart.

The western part of the Ethiopian Highlands is larger than the eastern part.

Geography of the Ethiopian Highlands

Around half of all of Ethiopia is covered by the Ethiopian Highlands, which hints at just how large this area truly is. In fact, the Highlands are so large some call it the "Ceiling of Africa". This area also contains one of the sources of the longest river in the world, the Nile. Lake Tana in the Highlands feeds a river known as the Blue Nile, which joins the other main tributary, the White Nile, in Sudan.

Formation

About 75 million years ago, part of the mantle rose upwards, pushing an enormous part of the Earth's crust up into a dome. Magma then fed volcanoes which ejected huge amounts of lava that built up the land further.

The rising mantle pushed upwards under the Earth's crust.

Mantle

Volcano

Eventually, the mantle forced the land up and volcanoes erupted onto the plateau.

Basalt is created when lava from volcanoes cools quickly. It is dark in colour.

Rocks

Many of the rocks found in the Ethiopian Highlands are igneous and formed by volcanic activity. Basalt and obsidian are two of these, and while they look different, both are formed by cooling lava.

Obsidian is volcanic glass, with sharp edges. It forms when lava cools without forming crystals.

Ancient volcanoes

Because of tectonic plate activity, this region of Africa has long been home to volcanoes. Several of these had earth-shaking eruptions between 320,000 and 170,000 years ago, creating peaks including the Simien Mountains in the north of the Highlands.

Arabian Plate

African Plate

Notice how the Red Sea runs nearly perfectly along the fault line between the separating plates.

Arabian Peninsula

The Arabian Peninsula used to be connected to the Ethiopian Highlands – that is, before the Arabian Plate pulled away from the African Plate. As the plates separated, water rushed in from the Indian Ocean to create the Red Sea. There is still a chain of mountains running along the west coast of the Arabian Peninsula.

Giant mole rat
(Tachyoryctes macrocephalus)

At weights of up to 2 kg (4 lb), giant mole rats definitely live up to their name. With big teeth and a love of digging, these rodents create complicated tunnel networks among the Highlands where they forage for plants.

Mountain nyala
(Tragelaphus buxtoni)

Mountain nyalas are related to antelopes. They are found in high altitude mountain woodlands, munching on leaves. Too big for Ethiopian wolves, the main predator of the nyala is the leopard.

Animals of the Ethiopian Highlands

When most people think about African wildlife, they immediately jump to elephants, giraffes, zebras, and crocodiles. The animals found in the Ethiopian Highlands are less well known, in fact, some of them are among the rarest animals on the planet. However, they are just as wondrous and adapted to a life at very high altitudes.

Walia ibex
(Capra wolie)

Both male and female walia ibexes have impressive horns. The males have larger horns than the females though, and also grow a long, black beard. Interestingly, this member of the goat family can often be found hanging out with geladas.

Thick-billed raven
(Corvus crassirostris)

If you think all ravens and crows look the same – think again! This one has a white patch on its head and the tip of its beak, as well as a huge bulge on its bill. These birds scavenge for food such as insect larvae and meat left by other predators.

Moorland chat
(Pinarochroa sordida)

While the moorland chat may look like a sparrow, these birds can be found sprinkled throughout the continent's highest areas. They can even be found at elevations up to 3,400 m (11,150 ft).

Ethiopian wolf
(Canis simensis)

With a red tinge to its fur and a pointy nose, the Ethiopian wolf looks a bit like a large red fox. This predator ventures far up into the Ethiopian Highlands – even to the highest point, Ras Dashen. Unfortunately, fewer than 500 of these wolves remain in the wild.

Gelada
(Theropithecus gelada)

Sometimes called "bleeding heart monkeys" due to the blood-red patch of skin on their chest, geladas live high up on clifftops where most predators can't go. They are grass-eaters that spend most of their time shuffling along on their bottoms grazing – they even have padded rears that act as cushions!

Adey abeba
(Bidens macroptera)

Adey abeba is a flower found only in Ethiopia, and it only appears for a short time each year. Yellow symbolizes peace, hope, and love to the people living in the Highlands, so they exchange bouquets of these flowers as gifts and to celebrate Ethiopian New Year.

African juniper
(Juniperus procera)

These trees can reach heights of up to 40 m (130 ft) and are the only species of juniper to be found south of the equator. The timber from the African juniper is used for making houses.

Plants of the Ethiopian Highlands

The Ethiopian Highlands' famous geladas wouldn't be able to survive without the plentiful grass found there, and they're not the only animals which depend on the area's native plants. Giant mole rats feast on roots and tubers, and walia ibex browse on the leaves of shrubs and bushes. For plants that can survive these hungry herbivores, the Highlands are an ideal place to grow, with cool temperatures and plenty of sunshine.

Tree heather
(Erica arborea)

Tree heather can grow taller than a giraffe and each spring erupts into clouds of dainty white flowers with purple centres. On some Ethiopian mountains, tree heathers grow in a strip from 3,000 m (9,840 ft) to 4,000 m (13,120 ft) up.

Coffee
(*Coffea arabica*)

It's hard to believe, but the dark, black coffee that adults drink comes from these tiny cherries, or fruit, from the coffee plant. This species is from the Ethiopian Highlands, but has now been taken to many different places across the globe.

Everlasting
(*Helichrysum citrispinum*)

This small shrub actually helps the plants growing beneath it. By providing shade and preventing water from flowing away, it creates its own mini habitat. It can also grow at very high altitudes – above 3,800 m (12,460 ft).

Torch lily
(*Kniphofia foliosa*)

Torch lilies, also known as red hot pokers, stand tall with magnificent flame-coloured flowers. This species is being investigated as a source of medicine for treating malaria.

African redwood
(*Hagenia abyssinica*)

Like its name suggests, African redwood has dark, red wood and it is actually a member of the rose family. This tree grows in cooler areas in the Ethiopian Highlands, usually above 2,000 m (6,560 ft).

Giant lobelia

At elevations of around 4,000 m (13,100 ft), you will find one of the weirder plants in the world. It's known as the giant lobelia, and it can reach more than 10 m (32 ft) in height. That's a flower spike that grows around three to five times as tall as a grown-up! Like the Queen of the Andes from South America, after this plant sends up its huge flower spike, it dies.

Lots of leaves

Giant lobelias are what's known as rosette plants, which means their leaves are arranged in a circle. At night, the leaves close inward, protecting the centre of the plant from the cold and forming what are called "night buds".

Taller at altitude

Scientists say that the giant lobelia's height is a bit of a superpower. This is because temperatures on the ground can vary. By growing tall, the giant lobelia actually rises above those cold temperatures and keeps itself warm.

Plant survivors

The fact that giant lobelia are often the tallest vegetation around shows that these curious plants thrive where other species can't even survive. Deep-growing roots and adaptations to the cold are key to the giants' success.

Stone forest

There aren't many places on Earth like the Sanetti Plateau in the Ethiopian Highlands. Years of erosion from wind, rain, and even glaciers carved these boulders and left them standing like trees turned to rock. If you stop and listen, you might just hear the howl of Ethiopian wolves echoing across the hills as wind whips through the stone forest.

Area
2,500,000 km²
(965,000 miles²)

Age
40 million years

Average height
4,500 m (14,700 ft)

Qing Zang Gao Yuan

With huge mountain ranges rising like walls to the north and south, Qing Zang Gao Yuan is a plateau like its own little world.

At an average height of 4,500 m (14,700 ft) above sea level, Qing Zang Gao Yuan is the largest and highest plateau on Earth. This region is also known for having a harsh climate, but it's home to thousands of species of plants and animals – and more than 10 million people. The area also spawns some of the longest and most important rivers on the planet.

Water source

One out of every six people on Earth drinks water that originally came from the glaciers, snowmelt, and weather on Qing Zang Gao Yuan. That's more than 1.3 billion people!

Surrounded by mountains

With Asia's longest mountain range, the Kunlun Shan, to the north and the Himalayas to the south, Qing Zang Gao Yuan is walled off from the rest of the continent.

THIS IS THE YOUNGEST, HIGHEST, AND LARGEST PLATEAU ON EARTH.

Qinghai Hu

Qing Zang Gao Yuan covers a whopping 2,500,000 km² (965,000 miles²).

Qinghai Hu

Like a miniature, inland ocean, Qinghai Hu is a huge, salty lake high on Qing Zang Gao Yuan. It provides a habitat for many migratory birds, and the endangered Przewalski's gazelle can be found near the lake.

Geography of Qing Zang Gao Yuan

The definition of a plateau is an area of relatively high, flat ground, and Qing Zang Gao Yuan definitely qualifies. This plateau is sandwiched between mountain ranges and is being pushed up by the same geological forces. It contains a variety of high-elevation habitats including grasslands, deserts, and steppes. Some call this region the Earth's "Third Pole", due to the huge amount of snow and ice it contains.

Formation

While scientists are still trying to figure out exactly how Qing Zang Gao Yuan formed, many believe it is the result of multiple tectonic events. This may explain why some areas have rock formations of different ages and origins.

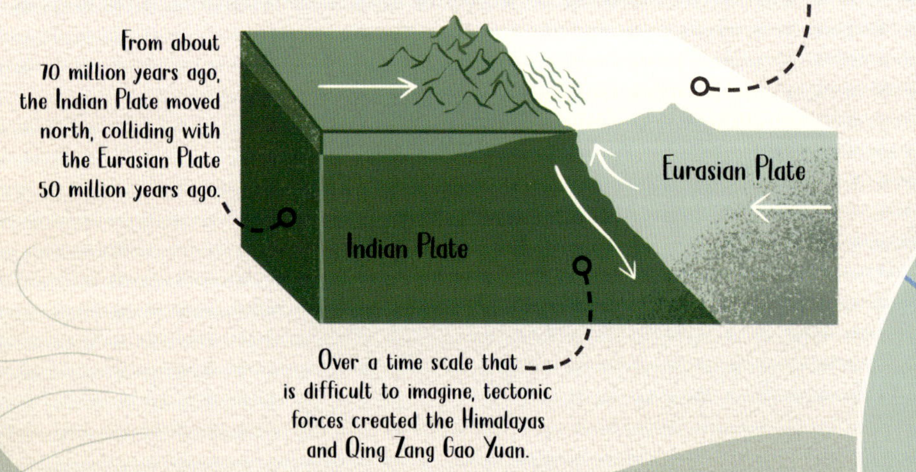

As the Indian Plate moved under, or was subducted beneath, the Eurasian Plate, the plateau and mountains were pushed up.

From about 70 million years ago, the Indian Plate moved north, colliding with the Eurasian Plate 50 million years ago.

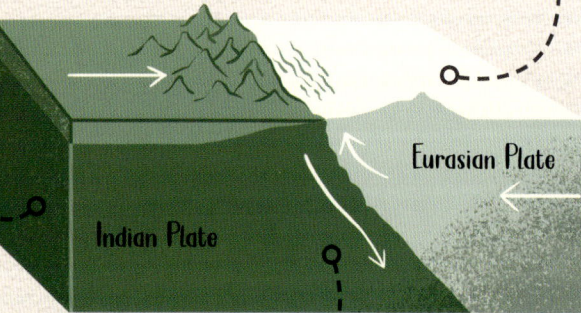

Eurasian Plate

Indian Plate

Over a time scale that is difficult to imagine, tectonic forces created the Himalayas and Qing Zang Gao Yuan.

River source

Like a giant funnel, Qing Zang Gao Yuan captures water from the highlands and directs it outwards into some of the largest rivers on the continent. Changjiang (the Yangtze), Lancang Jiang (the Mekong), Yarlung Zangbo Jiang, Huang He, the Salween, and the Indus are all rivers that begin somewhere on this plateau.

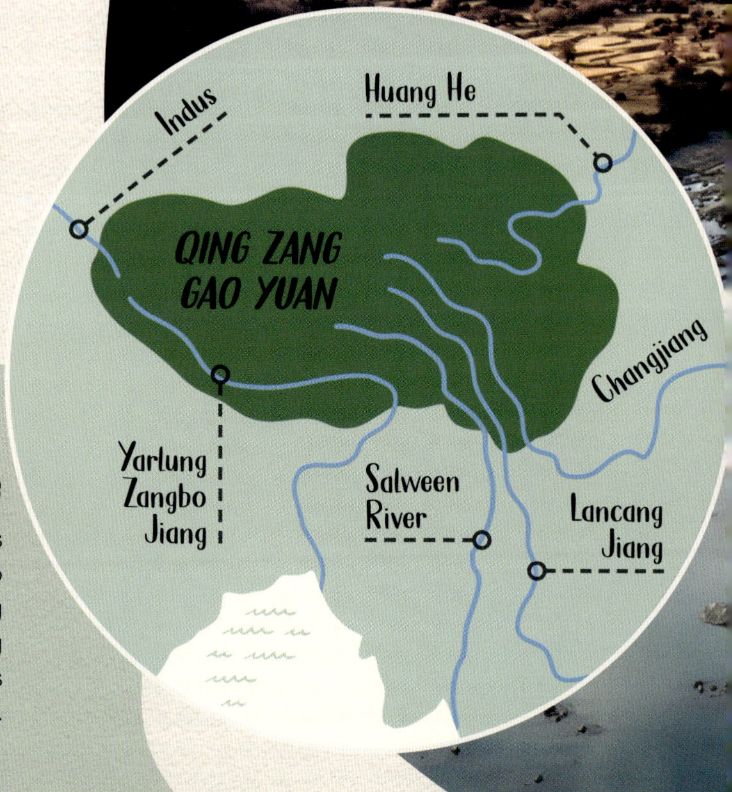

Indus

Huang He

QING ZANG GAO YUAN

Changjiang

Yarlung Zangbo Jiang

Salween River

Lancang Jiang

Monsoon season

While the Himalayas protect Qing Zang Gao Yuan from the worst of the yearly heavy rains, or monsoons, that strike India to the south, this region does receive more rain between late June and September. However, the summer rainy season does bring brilliant wildflowers!

Yarlung Zangbo Jiang Canyon

Apart from those in the oceans, the Yarlung Zangbo Jiang Canyon is the deepest canyon on Earth. This gorge also holds what some scientists believe is the planet's second tallest tree – a Himalayan cypress that stands more than 102 m (335 ft) tall.

Yarlung Zangbo Jiang

Hot spring snake
(Thermophis baileyi)

Reptiles don't usually do too well in cold climates, as they rely on the sun to warm up their blood. But these greyish-brown snakes have a trick – they get a boost from the warmth of geothermal pools!

Animals of Qing Zang Gao Yuan

While at first glimpse Qing Zang Gao Yuan may seem quite plain, lots of animals have evolved to survive here. In addition to the striking animals seen on these pages, this region is home to more than 2,300 species of insect, 64 species of fish, 45 species of amphibian, and 55 species of reptile!

Chiru
(Pantholops hodgsonii)

While lots of animals are poached for their horns, the chiru nearly went extinct because of its extremely soft underfur. When woven into yarn, this fur creates pieces of clothing that sell for up to £30,000 ($40,000) a piece!

Kiang
(Equus kiang)

Found in groups as small as five and as large as 400 animals, the kiang is a species of wild ass native to alpine grasslands. This species is also the largest wild ass species on Earth, with an average height taller than a grown-up.

Female chiru don't have horns. They gather together and migrate around 300 km (186 miles) to give birth to their babies.

Goa
(Procapra picticaudata)

The goa weighs up to 16 kg (35 lb), which means it's only as big as a medium-sized dog. Unfortunately, this species is threatened by hunting and habitat loss.

Black-necked crane
(Grus nigrocollis)

Even though this crane's name refers to the colour of its neck, the bright red "crown" of bare skin is what catches the eye. These long-legged birds can outrun humans.

Grey-sided fox
(Vulpes ferrilata)

With a long nose and a taste for rodents, the grey-sided fox is well-known for hunting in pairs and sharing kills. As such, these clever carnivores are important predators of the plateau, keeping populations of small mammals in check.

Keystone species

The plateau pika is a small but mighty mammal that the plateau ecosystem might just collapse without. By digging huge tunnel networks, pikas make homes for birds, lizards, insects, and plants. They also serve as a main food source for predators. Important animals like this are known as keystone species.

Wild yak

On Qing Zang Gao Yuan, you don't want to find yourself on the wrong end of a wild yak. They have been known to charge people who wander too close – and even vehicles! However, these relatives of the cow are usually gentle giants. In fact, yaks have been domesticated by people and are kept for their fur and milk.

Two-layered coat

With large hearts and super-charged blood with extra red blood cells, wild yaks are built for life on the cold, windy, and low-oxygen plateau. A thick, downy undercoat protects their core, while a layer of long, shaggy fur provides protection from the elements.

Herds

No one knows exactly how large the wild population of yaks is, but scientists estimate there are around 15,000 to 20,000 animals. Unfortunately for them, wild yaks eat the same sorts of grasses as domestic yaks and cattle, which sometimes gets them in trouble with local herders.

Strong horns

Wild yaks' horns can reach up to 98 cm (39 in) in length. Both males and females have horns, and a whole herd will sometimes form a circle and lower their horns to defend themselves and their young against wolves.

Plants of Qing Zang Gao Yuan

Qing Zang Gao Yuan has a reputation for being a cold desert, but it's actually home to a number of habitats, including forests, meadows, and steppes. Across those places, more than 5,800 species of plants sprout. Amazingly, there are even 17 different kinds of grassland here.

Cushion rock jasmine
(Androsace tapete)

Cushion rock jasmine is a curious-looking species and one of the highest-living flowering plants on the planet. Since it's adapted to life in the cold, scientists are worried that global warming could eventually cause it to go extinct.

Purple cone spruce
(Picea purpurea)

Yes, there really is a tree with purple pine cones! The bad news is that the purple cone spruce's wood is used to make things like furniture and musical instruments, and this has meant that it is now becoming rare.

Giant cowslip
(Primula florindae)

The giant cowslip can grow up to 1.5 m (5 ft) in height! While it's native to Qing Zang Gao Yuan, the plant's pretty, yellow flowers and ability to survive harsh climates has made it popular with gardeners everywhere.

Dwarf snow rhododendron
(Rhododendron nivale)

Like its name suggests, the dwarf snow rhododendron is on the small side when it comes to other rhododendrons in its family. It's also fond of cold habitats and can often be found growing right next door to glaciers.

Sichuan silk poppy
(Meconopsis punicea)

Closely related to the Himalayan blue poppy, the Sichuan silk poppy has dainty, downward-facing flowers that pop out a few at a time throughout the growing season. These plants can survive temperatures as low as –10°C (14°F).

Ancient cypress

Meet King Cypress, a 50 m- (164 ft-) high giant cypress tree. Found near the town of Bajie, some estimate it to be about 2,600 years old, which would mean it started growing when the ancient Egyptians lived.

Saltwater lake

With sapphire-blue waters and snow-capped peaks in the background, Nam Co is the highest saltwater lake in the world. It is found 4,718 m (15,479 ft) up on Qing Zang Gao Yuan and may be as deep as 120 m (394 ft). Despite being covered in ice for around half the year, there are two species of fish that live here.

Volcanoes

A volcano is an opening in the Earth's crust where melted rocks and incredibly hot gases can move upwards and spill onto the surface. Over time, so much rocky material can build up around these vents that entire mountains can form, such as Mount Fuji in Japan. Sometimes, volcanoes even help create islands, such as Mauna Loa in Hawaii.

Height
3,776 m (12,388 ft)

Age
11,000 years

Volcano type
Stratovolcano

Mount Fuji

Of all the lone mountains on Earth,
Mount Fuji may be the most
striking to behold.

Known as Fuji-san in Japanese, this large, picturesque
mountain towers over the island of Honshu. Mount Fuji
is an active volcano, with its last major eruption taking
place in 1707. It is also a place of spiritual and cultural
importance for Japanese people, and hundreds climb
the mountain each summer. Not only is Mount Fuji
a source of natural beauty and artistic inspiration,
but it has come to stand as a symbol of
Japan around the world.

Landmark

One of the country's three holy
mountains, and the tallest mountain
in Japan, Mount Fuji towers over the
lands surrounding it. This volcano
stands out as one of the most
recognizable landmarks
on the islands.

At their closest point, Japan and Russia are just 42 km (26 miles) apart.

Honshu is the largest island of Japan and Mount Fuji stands near its south coast.

Mount Fuji

Many volcanoes

The Japanese islands sit atop a region known for intense tectonic activity, and the country is currently home to 111 active volcanoes. That's about 9 per cent of all the active volcanoes on Earth.

From space

When viewed from space, astronauts tell us that many mountains look flat. But because Mount Fuji stands alone and is covered by snow at its highest elevations, there's no mistaking this volcano.

Geography of Mount Fuji

Mount Fuji is 125 km (78 miles) in circumference and has a diameter of around 45 km (28 miles). While this volcano only erupts about every 500 years, scientists still consider it to be active, not dormant. And since the last major eruption was in 1707, the next one should be sometime around the year 2200CE.

Ring of Fire

As much as 75 per cent of the planet's volcanoes, including Mount Fuji, can be found around the Ring of Fire – which is the name scientists give the chain of volcanoes that roughly outline the Pacific Ocean. It's not just volcanoes either. This zone is where 90 per cent of all known earthquakes occur, too.

Stratovolcano

With steep sides and a conical shape, Mount Fuji is what's known as a stratovolcano. Stratovolcanoes are sometimes called composite volcanoes, because they are made of layer upon layer of lava and volcanic ash laid down by many eruptions.

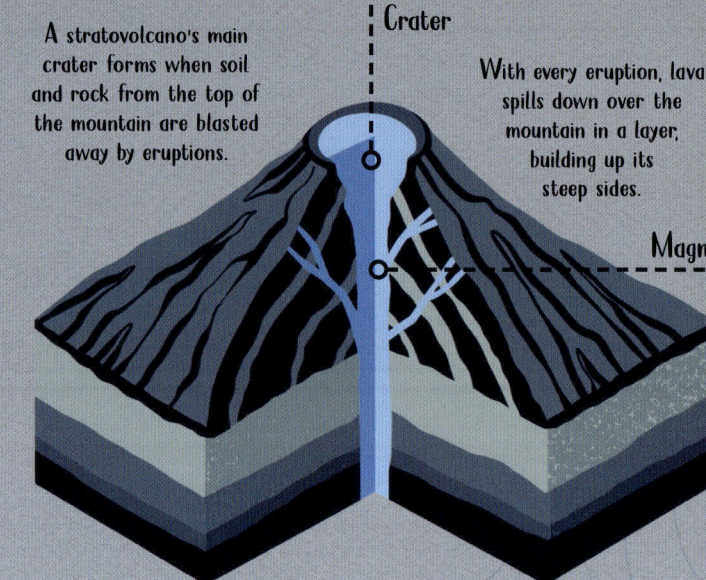

A stratovolcano's main crater forms when soil and rock from the top of the mountain are blasted away by eruptions.

Crater

With every eruption, lava spills down over the mountain in a layer, building up its steep sides.

Magm

When lava is still inside the Earth, it's known as magma. A magma chamber is where molten rock collects under a volcano before bubbling upwards.

EURASIAN PLATE

NORTH AMERICAN PLATE

PACIFIC PLATE

Tectonic plates move around Earth's surface very slowly. Their movement is what causes volcanoes and earthquakes.

AUSTRALIAN PLATE

NAZCA PLATE

Second crater

Sometimes volcanic ash and lava burst forth from an opening, or vent, on the side of a volcano, forming a second crater. In Mount Fuji's case, a second crater formed as recently as 1707.

New mountains can spring up as magma finds new ways to get to the planet's surface. And old mountains can disappear through erosion.

New Fuji

Old Fuji

Ashitaka

Komitake

Rocks of Mount Fuji

Pumice, andesite, and basalt are just a few of the kinds of rock Mount Fuji has been known to scatter during eruptions. Scientists can use volcanic rocks such as these to tell how old volcanoes are and how they've changed.

Mount Fuji through time

Volcanic mountains are a work in progress. About 400,000 years ago, before Mount Fuji came to be, escaping magma first formed another mountain known as Ashitaka. Some time later, yet another mountain called Komitake came to be. Finally, over the last hundred thousand years, Mount Fuji began to form.

Andesite makes up most of the inside of Komitake – a mountain now hidden inside Fuji.

Pumice is a volcanic rock that is full of holes. It is made from lava that contains bubbles of gas.

It's tough to imagine, but when basalt is molten, as it is in lava, it flows freely like a liquid.

Spotted nutcracker
(Nucifraga caryocatactes)

As a member of the corvid family, spotted nutcrackers are related to jays, crows, and ravens. Like their name suggests, these striking birds are both spotted and have a talent for hiding nuts that can be dug up and eaten in the winter.

Japanese pit viper
(Gloydius blomhoffii)

Known locally as the mamushi, this is one of the venomous snake species found in Japan. While mamushi only grow to a maximum length of 91 cm (36 in) and prefer to eat small rodents, accidental bites to people can be deadly.

Animals of Mount Fuji

One quarter of all the birds known to live in Japan have been sighted somewhere on Mount Fuji. Interestingly, different species can be found at different levels of elevation, with owls and warblers found down at the base and alpine accentors and spotted nutcrackers found way up beyond 2,500 m (8,000 ft). There are also freshwater fish, a handful of amphibians and reptiles, cicadas, dragonflies, and numerous species of butterflies.

Sika deer
(Cervus nippon)

Because the word "sika" means "deer" in Japanese, calling this animal a "sika deer" is actually repetitive. If you want to try a different name, you could call this animal a whistling deer, since that's the sound the males make when it's time to mate.

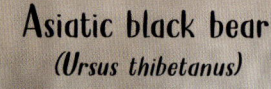

Asiatic black bear
(Ursus thibetanus)

Known for the cream-coloured 'V' or crescent moon-shape on its chest, the Asiatic black bear is native to Asia, from Iran all the way to the islands of Japan. While the bear's habitat is shrinking, its population in Japan is actually growing.

Alpine accentor
(Prunella collaris)

This house sparrow-sized bird thrives in bare mountain patches all over Europe and Asia, and Mount Fuji is no exception. Alpine accentors feast on spiders, snails, and worms and make small, cup-shaped nests out of grass, moss, and animal hair.

Warm coats

Some animals use warm coats of feathers or fur to help withstand the cold of the mountains. The Japanese macaque has thick, long hair for this purpose. It has also become world-famous for figuring out another luxurious way to stay warm – some go for a soak in volcanic hot springs!

Japanese serow
(Capricornis crispus)

The Japanese serow is a goat-like creature that can be found on Mount Fuji's slopes. It sometimes likes to stand on boulders and look out at the world all around. Other than humans and the occasional bear, serow have few predators in Japan.

Rhododendron
(*Rhododendron brachycarpum*)

Even when Mount Fuji is covered in thick snow, this rhododendron will still be green. As evergreens, rhododendrons keep their leaves all year instead of shedding them in the cold.

Fuji thistle
(*Cirsium purpuratum*)

If you were to hike the slopes of Mount Fuji between August and October, you wouldn't be able to miss the Fuji thistle. That's because at 1 m (3 ft) or more in height, it's the largest thistle in the country.

Japanese knotweed
(*Fallopia japonica*)

This plant protects itself against the Sun's harsh rays by producing chemicals called flavonoids, which act like sun cream. Interestingly, scientists have found that the plants that grow higher up on Mount Fuji's slopes make more chemicals than the plants down below.

Plants of Mount Fuji

As with Mount Fuji's wildlife, its plant life differs a great deal depending on the altitude. One reason for this is that the temperature drops about 0.6°C (1°F) with each 100 m- (328 ft-) increase in elevation. On the lower slopes, you might find firs, hemlocks, and beeches, but most tall-growing trees peter out around 2,500 m (8,200 ft). And at the highest elevations not covered with snow, moss and lichen rule.

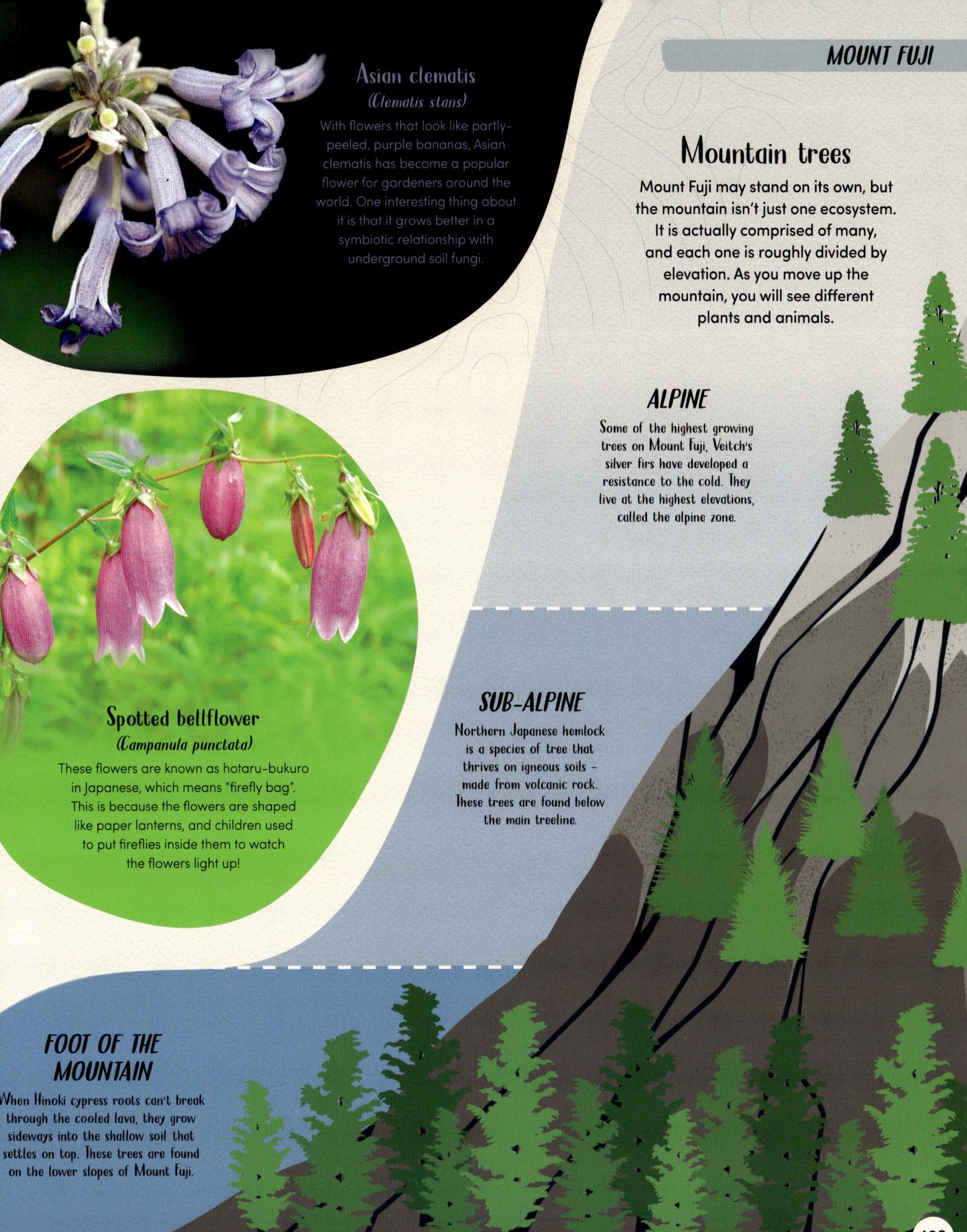

Asian clematis
(Clematis stans)

With flowers that look like partly-peeled, purple bananas, Asian clematis has become a popular flower for gardeners around the world. One interesting thing about it is that it grows better in a symbiotic relationship with underground soil fungi.

Mountain trees

Mount Fuji may stand on its own, but the mountain isn't just one ecosystem. It is actually comprised of many, and each one is roughly divided by elevation. As you move up the mountain, you will see different plants and animals.

ALPINE

Some of the highest growing trees on Mount Fuji, Veitch's silver firs have developed a resistance to the cold. They live at the highest elevations, called the alpine zone.

Spotted bellflower
(Campanula punctata)

These flowers are known as hotaru-bukuro in Japanese, which means "firefly bag". This is because the flowers are shaped like paper lanterns, and children used to put fireflies inside them to watch the flowers light up!

SUB-ALPINE

Northern Japanese hemlock is a species of tree that thrives on igneous soils – made from volcanic rock. These trees are found below the main treeline.

FOOT OF THE MOUNTAIN

When Hinoki cypress roots can't break through the cooled lava, they grow sideways into the shallow soil that settles on top. These trees are found on the lower slopes of Mount Fuji.

123

Fuji cherry

Every year, from late March to early April, Mount Fuji's foothills turn pink with the blossoms of Fuji cherry trees. Not only is this a glorious sight to behold, but the cherry blossoms represent good luck and hope to Japanese people, and their blossoming attracts tourists from all over the world.

Autumn colours

Each autumn, as the Fuji cherry prepares for winter, the chlorophyll that makes its leaves green starts to break down. This reveals yellows and oranges that were hidden by the green, as well as deep reds that appear as a result of a chemical reaction with sugars in the leaves.

Blossom

Fuji cherry trees erupt into colour as they blossom, but the display only lasts an average of 10 days. This is why the blossom, known in Japan as sakura, also symbolize impermanence – the blossom arrives, and then just as quickly, it is gone.

Flowers

People have been celebrating cherry blossoms in Japan for more than 1,000 years with a festival known as hanami. In Japanese, hana means "flower", and mi translates as "to view". Hanami is a time for gathering with family and friends to enjoy food and appreciate the natural beauty of cherry flowers.

Crater of Mount Fuji

With a depth of about 250 m (820 ft), Mount Fuji's main crater would be able to hold just over two-and-a-half Big Ben clock towers standing on top of each other. And even though this volcano remains active, as you can see in this photo the Okumiya Shrine sits at the mountain's summit – right at the edge of the crater below. The rocky pit is covered in pieces of a pumice-like stone known as scoria.

Height
4,169 m (13,677 ft)

Age
500,000 years

Volcano type
Shield volcano

Mauna Loa

With a slope that reaches all the way to the ocean's floor, Mauna Loa is the largest active volcano on Earth.

Mauna Loa is an active volcano on one of the islands of Hawaii, in the middle of the Pacific Ocean. Its name means "long mountain" in the Hawaiian language, referring to the way Mauna Loa stretches 10,210 m (33,500 ft) from its highest point all the way down to the bottom of the sea. Mauna Loa is sacred to many local people, who have been living on these islands for around 1,100 years.

Active volcano

Not only is Mauna Loa the world's largest volcano, but it's one of the most active on Earth. It has erupted 38 times since 1843, when modern records began.

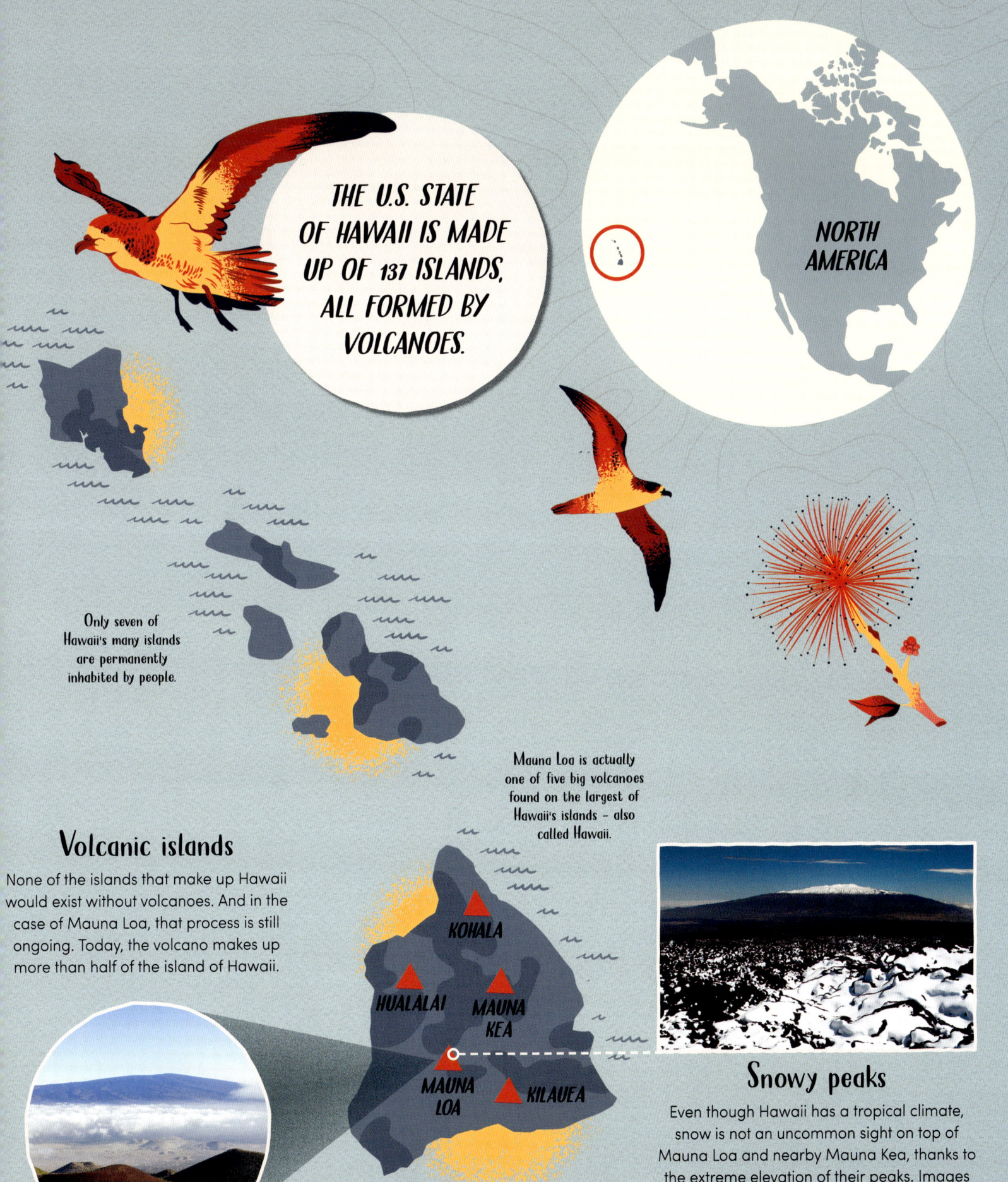

THE U.S. STATE OF HAWAII IS MADE UP OF 137 ISLANDS, ALL FORMED BY VOLCANOES.

NORTH AMERICA

Only seven of Hawaii's many islands are permanently inhabited by people.

Mauna Loa is actually one of five big volcanoes found on the largest of Hawaii's islands – also called Hawaii.

Volcanic islands

None of the islands that make up Hawaii would exist without volcanoes. And in the case of Mauna Loa, that process is still ongoing. Today, the volcano makes up more than half of the island of Hawaii.

KOHALA

HUALALAI

MAUNA KEA

MAUNA LOA

KILAUEA

Snowy peaks

Even though Hawaii has a tropical climate, snow is not an uncommon sight on top of Mauna Loa and nearby Mauna Kea, thanks to the extreme elevation of their peaks. Images from space show the two volcanoes as snow-capped thumbprints among an island of green.

129

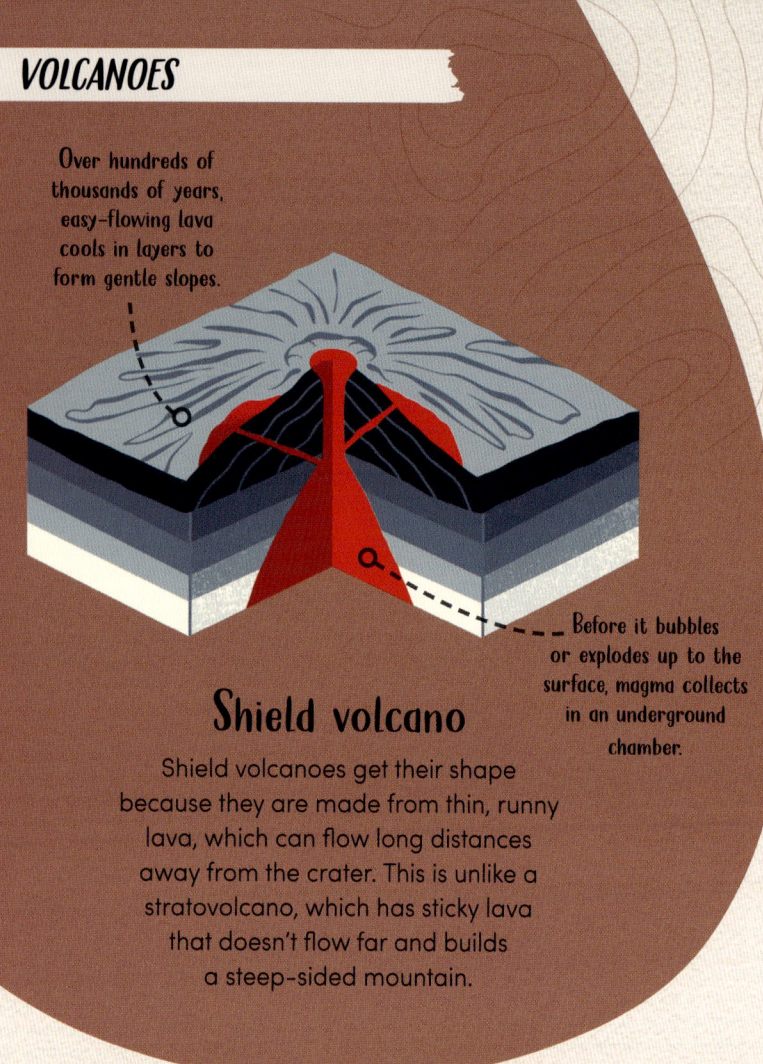

Over hundreds of thousands of years, easy-flowing lava cools in layers to form gentle slopes.

Before it bubbles or explodes up to the surface, magma collects in an underground chamber.

Shield volcano

Shield volcanoes get their shape because they are made from thin, runny lava, which can flow long distances away from the crater. This is unlike a stratovolcano, which has sticky lava that doesn't flow far and builds a steep-sided mountain.

Geography of Mauna Loa

Mauna Loa shares the island of Hawaii with four other volcanoes, but it is by far the biggest. If you were to hike up to the volcano's summit, you'd pass through lush tropical forests before arriving at fields of loose, jagged rock that might make it feel like you've landed on the surface of the Moon. Being an active volcano, Mauna Loa also releases volcanic gases, which can smell like rotten eggs.

Formation

A part of the mantle, called a plume, rose up over millions of years under the middle of the Pacific Plate. Some of it melted to form magma that escaped to the surface. As the plate above moved, the magma rose in a series of different spots, creating new islands in the Hawaiian chain.

In general, Hawaii's westernmost islands are the oldest, with the newest islands forming in the east.

Most volcanoes occur at the edge of tectonic plates, but those made by a mantle plume, like that under Hawaii, can happen in the middle of a plate.

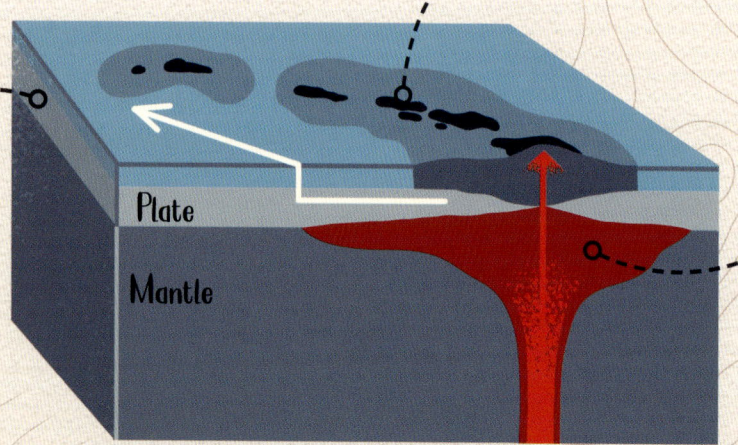

The mantle plume stays in roughly the same place, while the Pacific Plate slides over it.

Plate

Mantle

Runny lava

Depending on the combination of chemical elements inside it, some lava is thick and sticky, while other types are thin and runny. Mauna Loa's lava is more loose and runny which allows it to flow like a river of liquid rock down the mountain.

There are two forms of cooled lava on Mauna Loa – the first, called aa lava, is spiky and rough.

Highest mountain

Qomolangma Feng (Mount Everest) is widely known as the tallest mountain on Earth. But here's a secret – if you measured from the summit all the way down to where it actually begins – on the seafloor – Mauna Loa would be taller.

The second type of cooled lava on Mauna Loa looks a bit like rocky pancakes! It is smooth or wrinkly, and is known as pāhoehoe lava.

Lava tubes

Many lava flows cool to a thick crust on top, but magma keeps flowing for great distances underneath in tubes. When the magma drains away, the tunnels left behind look as if they were formed by colossal worms!

Lava cricket
(Caconemobius fori)

After a lava flow, lava crickets are one of the first living things to move back into the area. However, not much is known about these animals, and scientists aren't even sure where they live between eruptions!

Animals of Mauna Loa

Hawaiian hawk
(Buteo solitarius)

The Hawaiian hawk, or 'io, is the only hawk found on these islands. It's a small but mighty predator that feeds on other birds, as well as mammals that have been introduced to Hawaii by people, such as rats, mice, and mongooses.

It probably seems hard to believe that an active volcano has any wildlife at all, but as you'll see, it doesn't take long for animals to arrive after the lava has cooled. Because Mauna Loa is on an island, every species that lives here had to travel from across the sea, either on the waves or wind, or by being carried by birds. The exception of course is the animals that humans brought with them.

Hawaiian goose
(Branta sandvicensis)

The Hawaiian goose, or nēnē, probably once existed on each of the Hawaiian islands, but has become very rare since Europeans arrived. In fact, this bird almost went extinct in the 1950s, when just 30 geese remained.

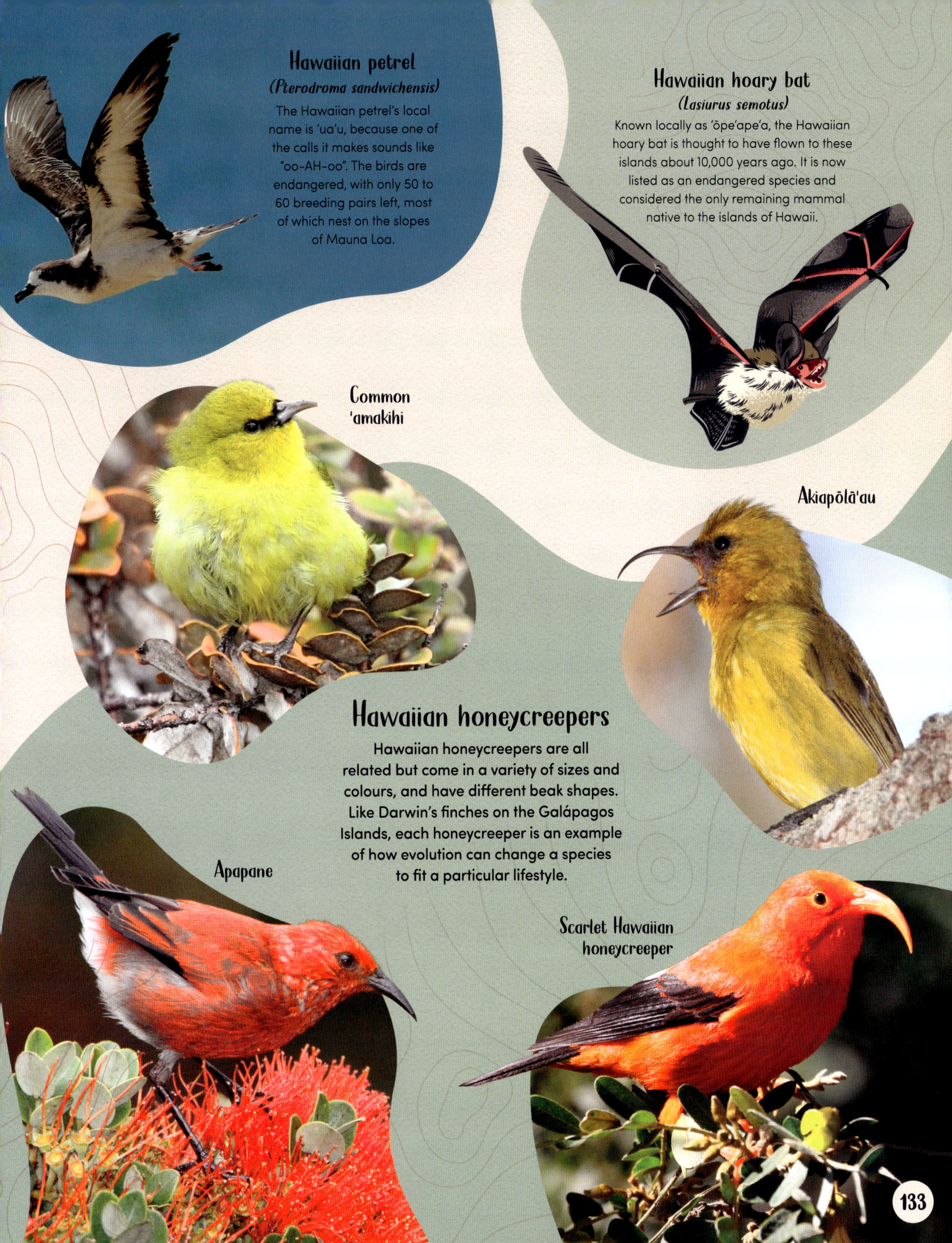

Hawaiian petrel
(*Pterodroma sandwichensis*)

The Hawaiian petrel's local name is 'ua'u, because one of the calls it makes sounds like "oo-AH-oo". The birds are endangered, with only 50 to 60 breeding pairs left, most of which nest on the slopes of Mauna Loa.

Hawaiian hoary bat
(*Lasiurus semotus*)

Known locally as 'ōpe'ape'a, the Hawaiian hoary bat is thought to have flown to these islands about 10,000 years ago. It is now listed as an endangered species and considered the only remaining mammal native to the islands of Hawaii.

Common 'amakihi

Akiapōlā'au

Hawaiian honeycreepers

Hawaiian honeycreepers are all related but come in a variety of sizes and colours, and have different beak shapes. Like Darwin's finches on the Galápagos Islands, each honeycreeper is an example of how evolution can change a species to fit a particular lifestyle.

Apapane

Scarlet Hawaiian honeycreeper

ʻŌhelo ʻai
(Vaccinium reticulatum)

ʻŌhelo ʻai produces bright red berries. A threatened goose, known as the nēnē, relies on these berries for food, as do several species of moths. The plant can be found growing on recent lava flows.

Māmane
(Sophora chrysophylla)

Lei wreaths are garlands of flowers that are popular gifts in Hawaii. The yellow blooms come from māmane trees. This tree was also once prized for its wood, which was used to make houses, tools, and provided fuel for fires.

Plants of Mauna Loa

Since Hawaii is in the middle of an ocean, all of its plants had to find their way to these islands in ingenious ways – on the waves, in the bellies or on the feathers of birds, or even on the wind. And once they got there, they evolved into unique species, about 90 per cent of which can't be found anywhere else.

Koa bean pods

Koa
(Acacia koa)

Koa is Hawaii's tallest native tree, with the ability to grow upwards of 33 m (115 ft). Historically, koa wood was reserved for Hawaiian royalty, and it remains one of the most valued hardwoods on the planet.

'A'ali'i
(Dodonaea viscosa)

After lava flows, 'a'ali'i is one of the first pioneer plants to pop back up, adding a dash of colour to a world turned black. This species is extremely tough, able to withstand great gusts of wind and grow at elevations up to 2,400 m (8,000 ft).

Pūkiawe
(Styphelia tameiameiae)

Like 'ōhelo 'ai berries, the small red fruit of pūkiawe are a popular food for nēnē geese. Although the geese love them, the berries are poisonous to humans! This plant can grow as a shrub or a tree.

Mauna Loa silversword
(Argyroxiphium kauense)

This impressive plant, which can grow more than 3 m (9 ft) in height, gets its name from its long, sword-like leaves and the silver, hairy structures that protect them. Interestingly, this species flowers only once in its 30-year life cycle, and the whole plant then dies.

135

'Ōhi'a lehua

'Ōhi'a lehua is such a common tree on the islands of Hawaii that up to 80 per cent of the forests there are made up of it. This large, slow-growing tree, like many of the other successful species found in Hawaii, is really good at growing on top of recent lava flows. This is thanks to its deep, strong roots, which also help protect against erosion and flooding by holding loose soil in place.

Size difference

Travel across the islands and you may notice that 'ōhi'a lehua trees come in lots of different shapes and sizes – from crooked, little shrubs to towering trees of up to 25 m (82 ft). Higher up, where it grows directly on cooled lava, this tree is usually smaller.

Leaves

Plants breathe through stomata, or pores, on their leaves. But 'ōhi'a lehua leaves have a clever trick – they can actually close those pores, sort of like holding their breath, to prevent damage from toxic volcanic gases.

Flowers

Almost like the trees are trying to imitate the colours of fire and lava, 'ōhi'a lehua blooms into a blaze of reds, pinks, yellows, and even whites. The long centres of the flowers might attract honeycreepers to pollinate them.

Recent eruption

As the largest active volcano in the world, perhaps it's not so surprising that Mauna Loa has erupted as recently as November 2022. That was the first time it had erupted since 1984, however. Fire fountains from giant cracks near the summit fed lava flows, but fortunately nobody was injured and the eruption didn't affect any heavily populated areas.

Tepuis

Tepuis are like a cross between a mountain and a plateau. Also known as tabletop mountains, the tepuis of South America are what's left of a giant plateau that existed 1.7 billion years ago, but which has since partially worn away. Many tepuis, including Mount Roraima, stand apart from each other in a sea of rainforest.

Height
2,810 m (9,220 ft)

Age
70 million years

Area
31 km²
(12 miles²)

Mount Roraima

Where the countries of Brazil, Venezuela, and Guyana meet, there's a cloud-shrouded island in the sky known as Mount Roraima.

Standing on Mount Roraima, it might seem as though you're looking over the edge of the world. Located in the north of South America, Mount Roraima is one of a string of flat-topped mountains known as tepuis. Each tepui is a wondrous mix of jungle, cliffs, waterfalls, and sky. Rising up out of the rainforests below, many animals and plants were stranded on the tepuis as they formed, setting each on a unique evolutionary path. This makes the species on every tepui not just different from those in the forest below, but from each other.

Mini plateau

Believe it or not, but what is now an area of high-in-the-sky plateaus was formed at around sea level. Today, this formation stretches across numerous countries and is known as the Guiana Shield

Mount Roraima is one of many tepuis that are part of the Guiana Shield.

The Guiana Shield formation covers roughly 1.5 million km² (600,000 miles²) in northern South America.

Cloud blanket

One of the factors that makes Mount Roraima and the rest of the tepuis so stunning is the way they seem to rise from the clouds. This happens when cool, moist air is trapped in the valleys around them, forming thick clouds that hide the forest below.

Viewed from above, Mount Roraima and the other tepuis can look like islands in a sea of cloud.

Mount Roraima

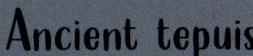

Mount Roraima is found in the Pacaraima Mountains, which make up part of the Guiana Shield.

Ancient tepuis

The word tepui comes from the local Pemón language and means "house of the gods." The tepuis are the remains of an old sandstone plateau that existed in this area, and are some of the most ancient landforms on the continent.

THE ROCKS OF SOUTH AMERICA'S TEPUIS ARE AROUND 2 BILLION YEARS OLD.

Formation

Mount Roraima is made of a rock called sandstone. For more than a billion years, all the tepuis were part of a much larger sandstone plateau called the Guiana Shield. Through the ages, the sandstone was ground away by erosion, leaving behind just isolated chunks, which are now known as the Pacaraima Mountains.

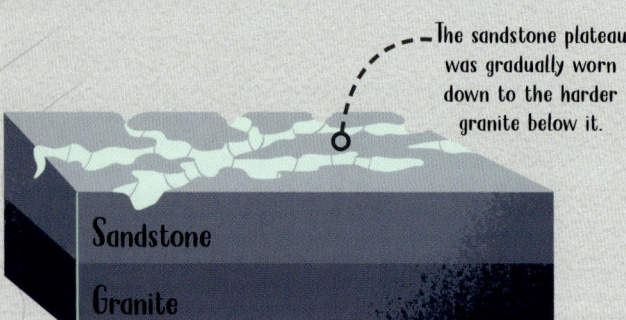

The sandstone plateau was gradually worn down to the harder granite below it.

Sandstone

Granite

Tepuis tend to stand alone, unlike mountain chains in other mountain ranges.

Between the tepuis is lowland forest habitat, completely different from what's above.

Geography of Mount Roraima

Mount Roraima is a well-known high point within the larger Pacaraima Mountains of South America. Unlike pointy pinnacles such as the Matterhorn, Mount Roraima is flat on top, like a table, which really gives it the illusion of being another world lifted up into the sky. While rainforest reigns below, smaller plants grow up here, and where water meets the edge, it creates spectacular waterfalls.

Rock formations

Strange rock formations stand guard like faceless gargoyles on the top of Mount Roraima. Each was formed as erosion by wind and rain ate into the sandstone rocks, similar to how the tepui they stand upon was created.

Known locally as La Tortuga, this famous rock looks like a sea turtle that has been turned to stone.

Colourful caves

Erosion also means the tepuis contain a vast network of caves, but these are unlike any others. Pink quartz gives the walls a rosy glow, and chemicals in the streams can turn the water red! Scientists have even found a mineral in the tepuis not seen anywhere else – it's called rossiantonite.

Valley of Crystals

Grains of the mineral quartz form the tepui sandstones. Huge hexagonal crystals of quartz spill out into the Valley of Crystals on Mount Roraima. To make sure the valley remains intact for everyone to enjoy, taking souvenirs is not only forbidden – it's considered bad luck!

Quartz is most often clear or white, but if manganese or iron are present, it can also be pink!

145

Roraiman rocket frog
(Anomaloglossus roraima)

When most people think of frogs, they picture ponds and streams. But the Roraiman rocket frog breeds in a plant known as a bromeliad. With hollow centres, bromeliads collect rainwater and act like mini freshwater habitats!

Animals of Mount Roraima

Scientists have wondered for a long time how the animals on top of the tepuis got there. Were they already there when the mountains rose, or did they climb or hitchhike their way to the top? Birds can fly, of course, but what about frogs? 87 per cent of those found on the tepuis don't exist anywhere else. While we still aren't quite sure, the animals found here are truly unique.

Roraima butterfly
(Pedaliodes roraimae)

This butterfly was described more than 100 years ago in 1912, and still, not much is known about its behaviour or lifestyle. What we do know is that is closely related to many other butterfly species in the region.

Roraiman mouse
(Podoxymys roraimae)

Scientists know almost nothing about some of Mount Roraima's wildlife. For instance, no one can say how long the Roraiman mouse lives, how it reproduces, or what predators it has. One thing we can say? It likes to eat earthworms.

Roraima bush toad
(Oreophrynella quelchii)

With black, bumpy skin and large eyes, this toad seems like it might be nocturnal. However, it's actually active during the day and can be found climbing on rocks near the summit of Mount Roraima.

Greater flowerpiercer

Roraiman barbtail

Tepui wren

Birds of Mount Roraima

While many other animals cannot easily find their way up onto Mount Roraima, birds are the exception. More than 200 species have been documented up here, including several species of owls, hawks, parrots, and woodpeckers. It is more common to see small perching birds though.

147

Oilbird

Meet the only flying, nocturnal, fruit-eating bird in the world! These birds get their strange name because local peoples used to catch their chubby chicks and use them as a source of oil for flavouring food and lighting torches. They are found in many habitats around northern South America, including mountains at elevations of more than 3,000 m (9,840 ft).

Cave roosts

Oilbirds make their nests in caves, including those found on tepuis. In Spanish, the oilbird is known as the guácharo, which means, "the one who whines". This is due to the raspy wail it makes while roosting amongst other oilbirds.

Echolocation

Like some bats, oilbirds use echolocation to avoid flying into things in the dark. They make loud clicks and listen for the returning echoes to find out if there are any objects nearby. These curious birds also have long whiskers next to their beaks which help them feel around in the dark.

Nests

Oilbirds build their nests on ledges inside caves using a mixture of regurgitated fruit and their own droppings. What if people built nurseries for their babies out of vomit and poo? Oilbird chicks don't seem to mind though.

Plants of Mount Roraima

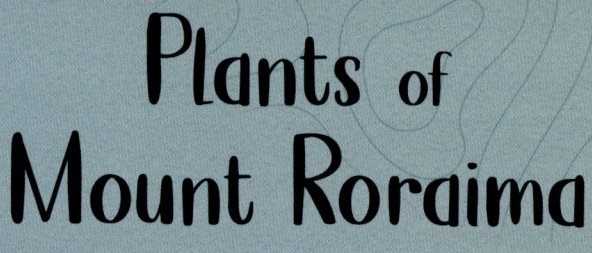

As with animals, many of the plants you'll find on top of the tepuis are endemic – which means they're found here and nowhere else in the world. In fact, scientists think up to 60 per cent of the plant life here is unique to South America's sky islands. Curiously, there are fewer pollinators way up top, which leads the plants to bloom longer than normal.

Roraima sundew
(Drosera roraimae)

Unlike slippery pitcher plants, the Roraima sundew catches its victims by being sticky. Red tendrils lure in mosquitoes and other insects and then trap them in blobs of glue – just like fly paper!

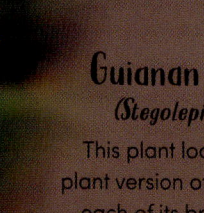

Guianan stegolepis
(Stegolepis guianensis)

This plant looks a bit like the plant version of a sea urchin, but each of its brightly-coloured spikes is actually a flower head. Each pointed bud opens into a yellow bloom – the one in this picture is just starting to open!

Roraima bonnetia
(Bonettia roraimae)

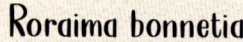

This plant is interesting in that it can be small, like a shrub, or tall, like a tree. On the top of tepuis, the Roraima bonnetia only reaches small heights, but it stands tall over other tepui plants.

Pitcher plant
(Brocchinia reducta)

Did you know some plants eat meat? Pitcher plants hold digestive juices in their centre, which break down any insects or small animals that fall in. The prey provide nutrients to the plant.

Poor soil

There's very little soil on top of the tepuis, as most of what used to be here has been carried away by wind and water over the ages. What's left is low in nutrients, which affects what kinds of plants can live here. It is also why so many tepui plants are carnivorous – they get the nutrients they need from their prey.

Marsh pitcher plant
(Heliamphora nutans)

This carnivorous plant has a special trick. It has hairs that point down into its pitchers that get super slippery when wet, creating a sort of botanical water slide! Scientists found that ants are three times more likely to be caught when the plants are wet.

Roraima azalea
(Bejaria imthurnii)

Although not actually an azalea, this plant is related to the low-growing shrubs. This species grows on the rocky surface of Mount Roraima and has delicate pink flowers and a dense covering of leaves on its stems.

Sceptre orectanthe
(Orectanthe sceptrum)

This plant is found dotted about the landscape of Mount Roraima. It looks a bit like the top of a pineapple, and like the pineapple plant, it can only flower once. It sends up a long stalk with a yellow flower at the end, but afterwards it dies.

Pink sand

The rock formations on top of Mount Roraima have been weathered to black or dark in colour, but patches of sand are salmon pink! This happens when the rock breaks down, revealing the true colours of the quartz grains – yes, the very same stuff that makes up the Valley of Crystals! These quartz grains are a mixture of red, white, and pink, so the sand is a pink colour, too.

Glossary

ADAPTATION
characteristic of a living thing that has developed over many generations to help it survive better in its environment, such as dense fur or a long tail

ALPINE
description of something relating to high mountains, especially the Alps in southern Europe

ALTITUDE
measurement of how high something is on land or in the air. It is usually measured from the ground or from sea level

ATMOSPHERE
layer of gases found above Earth's surface that separate it from space. The atmosphere contains the oxygen that organisms need to breathe and traps heat from the Sun, keeping the Earth warm

CARNIVORE
organism that eats meat. Animals that are carnivores are often predators

CONIFEROUS
description of plants that produce seeds in cones. Many conifers have needle-shaped leaves and are evergreen, keeping their leaves all year round

CORE
innermost part of the Earth, thought to be roughly spherical and composed of a solid metal inner sphere with a liquid metal outer layer

CLIMATE
long-term pattern of weather in a particular area

CROP
plant that is grown for food, material, or fuel, such as wheat grown to make flour

CRUST
rocky, outermost layer of the Earth, on which humans and all other life forms exist. The crust and part of the mantle below are broken into tectonic plates

DECIDUOUS
description of plants that shed their leaves in the autumn or during the dry season

DESERT
habitat characterized by having very little rainfall and sparse plant life. Some deserts are hot, but others are cold

ECOSYSTEM
network of living things and the environment to which they are adapted, including the climate in that area

ELEMENT
any one of the basic substances that make up all matter on Earth and which cannot be broken down into other substances

ELEVATION
measurement of how high something is on land. It is usually measured from the ground or from sea level

ENDANGERED
description of a species with few remaining individuals, that is in danger of becoming extinct

EROSION
geological process by which rock or soil is worn away by the action of wind, ice, water, and other natural phenomena

EVOLUTION
process by which species change slowly over generations, often becoming better suited to their environment

EXTINCT
description of a species that no longer exists

GEOLOGY
science of rocks and minerals as well as the structure of the Earth and how it has changed over time

GEOTHERMAL
description of heat produced by the interior of the Earth

GLACIER
huge mass of ice that lasts for centuries, and which moves slowly down mountain valleys

HABITAT
area that is home to a plant or animal. Different habitats have different climates and plant coverage. For example, forests, mountains, grasslands, and deserts are all types of habitat

HIBERNATION
when a living thing goes into a sleep-like state in order to survive periods when food is scarce or the weather is harsh

IGNEOUS
type of rock created by the cooling of magma or lava

HERBIVORE
organism that eats plants. Animals that are herbivores are often prey

LAVA
hot, liquid rock that flows onto the Earth's surface. Also the rock formed when it cools

MAGMA
hot, liquid rock that is found deep below the Earth's surface

MANTLE

largest layer of the Earth's interior that is sandwiched between the core and the crust. The mantle is made from hot rock

METAMORPHIC

type of rock that is made from other kinds of rock that have been changed by extreme heat or pressure

MIGRATION

when a species moves from one place to another to find better food supplies or to raise young, typically every year

MINERAL

solid substance that is made from elements. Rocks are made from minerals

MOUNTAIN

landform that is high in elevation compared to the surrounding land and often has a steep, rocky peak

MOUNTAIN RANGE

series of mountains that are close together, often found in a long line along a plate boundary

NOCTURNAL

description of a living thing that is active at night

PLATEAU

landform that is high in elevation compared to the surrounding land and which is mostly flat on top

POLLINATION

process by which flowering plants reproduce, where pollen is transferred from one plant to another either by the wind or by animals, known as pollinators

PRECIPITATION

process by which a form of water falls from the clouds onto the ground as weather. For example, water droplets fall as rain, or ice crystals fall as hail or snow

RAINFOREST

habitat characterized by high levels of rainfall throughout the year. Rainforests are usually full of different plants and animals

RODENT

group of mammals that have long, sharp front teeth that grow continuously

SANSKRIT

ancient language from southern Asia

SATELLITE

machine that is sent into space and which orbits a planet or other object in order to observe it

SEA LEVEL

average height of the surface of the ocean. It is often used as a starting point for measuring height or depth

SEDIMENT

grains of solid material, such as rocks or minerals, often created by erosion. Sediment can be washed into new locations by rain and rivers

SEDIMENTARY

type of rock that is made from layers of sediment that have been compacted over time until the grains stick together

SCREE

mass of rock fragments often found on the side or at the foot of a mountain or hill

SPECIES

group of living things that look similar and which can reproduce together

SNOW LINE

imaginary line on a mountain or other area of high elevation above which some permanent snow remains frozen throughout the year

STEPPE

habitat characterized by grassy plains with a temperate climate

SUBTROPICAL

description of a region that is warm and humid, found between tropical and temperate regions

TECTONIC PLATE

one of several gigantic slabs of the Earth's surface made up of a piece of the crust and the upper part of the mantle below it. Tectonic plates move around slowly which can create mountains, volcanoes, and other landforms

TEPUI

kind of tabletop mountain found in South America

TEMPERATE

description of a region with moderate temperatures for most of the year

THREATENED

description of a species that has declined in numbers and is at risk of becoming endangered

TREE LINE

imaginary line on a mountain or other area of high elevation above which trees cannot grow because of cold temperatures. May also be known as the timberline

TROPICAL

description of a region that is hot and humid for most of the year

TUNDRA

habitat characterized by cold temperatures and no trees. This type of habitat is found around the Arctic and on mountains

ULTRAVIOLET RADIATION

invisible rays that are part of the energy emitted by the Sun

VOLCANO

opening in the Earth's surface where lava and hot gases erupt onto the land

Index

Acknowledgements

DK would like to thank the following people for their assistance in the preparation of this book: Jonathan Melmoth for proofreading; Rituraj Singh for picture research assistance; Helen Peters for the index; and Simon Mumford for cartography.

The publisher would like to thank the following for their kind permission to reproduce their photographs:

(Key: a-above; b-below/bottom; c-centre; f-far; l-left; r-right; t-top)

1-160 Dreamstime.com: Jackreznor (All pages-Background). 4–5 Alamy Stock Photo: photomalcolm / Stockimo. 6–7 Alamy Stock Photo: Lisandro Trarbach. 8–9 Alamy Stock Photo: Eyal Bartov. 10–11 Alamy Stock Photo: David Dorey - Tanzania Collection. 12–13 Alamy Stock Photo: Lens And Light / Balan Madhavan. 14–15 Alamy Stock Photo: imageBROKER.com GmbH & Co. KG / Robert Haasmann. 17 Dreamstime.com: Ivan Kluciar (bl); Mikael Males (br). Science Photo Library: Dimitri Weber / Amazing Aerial Agency (t). 19 Alamy Stock Photo: David R. Frazier Photolibrary, Inc. (b). 22 Dorling Kindersley: University Museum of Natural History (cl, ca, clb, cra). 23 Alamy Stock Photo: Gresko81 / RooM the Agency (b); Roberto Moiola (tr). 24 Depositphotos Inc: MarkCauntPhotography (bc). NASA: (cra, cr). 25 Alamy Stock Photo: Mikel Bilbao Gorostiaga- Nature & Landscapes (t); Nick Yates / StockShot (b); Rowan Morgan (cra). 26–27 Alamy Stock Photo: imagoDens. 29 Getty Images / iStock: Igor Alecsander (crb). NASA: JSC (cl). 30–31 Alamy Stock Photo: David South (bc). 31 Alamy Stock Photo: Siim Sepp (cr). Dreamstime.com: Sara Winter (t). 32 Alamy Stock Photo: John Holmes (cl); Pep Roig (br); imageBROKER.com GmbH & Co. KG / C. Huetter (bl); Wayne Lynch / All Canada Photos (tr); Octavio Campos Salles (cr). 33 Dreamstime.com: Delstudio. 34 Alamy Stock Photo: James Brunker (bl); stefanophotographer (tc); Emmanuel LATTES (br). Dreamstime.com: Jose Ignacio Naranjo (tl). 35 Alamy Stock Photo: Florian Kopp / imageBROKER.com

GmbH & Co. KG (tc); David Vilaplana (br). **Dreamstime. com:** Irina Borsuchenko (cl). **36 Alamy Stock Photo:** Zoonar GmbH (tr). **36-37 Alamy Stock Photo:** Cyril Ruoso / Nature Picture Library. **37 Alamy Stock Photo:** Cyril Ruoso / Nature Picture Library (ca). **Dreamstime. com:** Jonathan Chancasana (bc). **38-39 Alamy Stock Photo:** Zenobillis. **41 Alamy Stock Photo:** imageBROKER.com GmbH & Co. KG / Wigbert Rth (tc). **Science Photo Library:** Mark Williamson (br). **42 Dreamstime.com:** Ursula Perreten. **43 Alamy Stock Photo:** The Natural History Museum (cl). **Science Photo Library:** Natural History Museum, London (ca). **44 Alamy Stock Photo:** Emil Oprisa (bl); Martin Sneary (tl); Alex Treadway / robertharding (tr). **naturepl.com:** Gavin Maxwell (cr). **45 Alamy Stock Photo:** Nilanjan Chatterjee (tr); JSK (cl). **46 Alamy Stock Photo:** Andrew Cline (c). **46-47 Dreamstime.com:** Ondej Prosick. **47 Alamy Stock Photo:** imageBROKER.com GmbH & Co. KG / Vikram Singh (tr). **naturepl.com:** Nick Garbutt (br). **48 Dreamstime.com:** Max5128 (t). **49 Alamy Stock Photo:** Bob Gibbons (tl); John Richmond (cl). **naturepl. com:** Dong Lei (cr). **Shutterstock.com:** Jiang Tianmu (tr, br). **50-51 AWL Images:** ClickAlps. **52 Alamy Stock Photo:** GFC Collection RF (cra). **53 Alamy Stock Photo:** Claudia Weinmann (tr). **54 Science Photo Library:** Dr Juerg Alean (cr). **54-55 Science Photo Library:** Bernhard Edmaier (t). **56 Alamy Stock Photo:** blickwinkel / S Gerth (tl); INTERFOTO / Zoology (bl). **56-57 Alamy Stock Photo:** Bernd Zoller / imageBROKER.com GmbH & Co. KG (bc); Tierfotoagentur / M. Zindl (tc). **57 Alamy Stock Photo:** Arco / J. Fieber / Imagebroker (cl); Arndt, S.-E. / juniors@wildlife / Juniors Bildarchiv GmbH (br). **58 Alamy Stock Photo:** Giulio Ercolani (br); Marcel Gross (l). **58-59 Alamy Stock Photo:** Reinhard Hlzl / imageBROKER.com GmbH & Co. KG (t). **59 Alamy Stock Photo:** Dave Derbis / mauritius images GmbH (br); McPhoto / Volz (cl); Paul Harcourt Davies / Nature Picture Library (cr). **60 Alamy Stock Photo:** Fotofeeling / Westend61 GmbH (cl). **60-61 Dreamstime.com:** Ivan Kluciar. **61 Alamy Stock Photo:** Botanicum (cra); Alexey Senin (crb). **62-63 Alamy Stock Photo:** Joana Kruse. **65 Getty Images / iStock:** Stockbyte (cla). **Science Photo Library:** Phil Degginger (bl). **66 Alamy Stock Photo:** George Ostertag (b). **67 Alamy Stock Photo:** Jason Bazzano (cr); Kevin Schafer (clb). **68 Alamy Stock Photo:** Bruce Montagne / Dembinsky Photo Associates (cr); Gabriel Rojo / Nature Picture Library (tl); Jared

Hobbs / All Canada Photos (b). **naturepl.com:** Charlie Summers (tr). **69 Alamy Stock Photo:** Dubi Shapiro / AGAMI Photo Agency (cl); Nick Trehearne / All Canada Photos (b). **70 Alamy Stock Photo:** Roberta Olenick / All Canada Photos (tl); Sumio Harada / Minden Pictures (cl). **70-71 Alamy Stock Photo:** Andrey Podkorytov. **71 Alamy Stock Photo:** Sumio Harada / Minden Pictures (crb). **72 Alamy Stock Photo:** Bob Gibbons (br); Adam Schneider (tl). **Getty Images:** Federica Grassi (bl). **73 Alamy Stock Photo:** Bob Gibbons (tl); Bryan Reynolds (tr); Pearl Bucknall RF (c). **74-75 Dreamstime. com:** Timon Schneider. **77 Alamy Stock Photo:** Viktor Posnov (bc). **Getty Images / iStock:** Sezgin Aygun (t). **78-79 Alamy Stock Photo:** Ben Goode. **79 Alamy Stock Photo:** geoz (bc). **Dreamstime.com:** Bjrn Wylezich (cb). **80 Getty Images:** Auscape / Universal Images Group (bl); Jason Edwards (t). **80-81 Alamy Stock Photo:** Selfwood (bc). **81 Alamy Stock Photo:** Rick & Nora Bowers (br). **Rudie H. Kuiter:** (cl). **82-83 Dreamstime. com:** Ken Griffiths. **82 Dreamstime.com:** Ken Griffiths (cb). **83 Getty Images:** Bill Blair (ProAIRvision) (br); The Sydney Morning Herald / Peter Rae / Fairfax Media (tr). **84 Alamy Stock Photo:** Frederic Tournay / Biosphoto (tr/new); Manfred Gottschalk (b). **85 Dreamstime.com:** Adrian Eugen Ciobaniuc (cl). **Joe McAuliffe:** (br). **86-87 Dreamstime.com:** Taras Vyshnya. **88-89 Alamy Stock Photo:** Wirestock, Inc.. **91 Getty Images:** Emad aljumah (br). **92-93 Alamy Stock Photo:** Christian Offenberg (c). **93 Alamy Stock Photo:** Clement Philippe / Arterra Picture Library (tc). **Dreamstime.com:** Onlyfabrizio (ca). **94 Alamy Stock Photo:** Martin Lindsay (b). **naturepl.com:** Sylvain Cordier (tl). **95 Alamy Stock Photo:** Jez Bennett (br); Justine Pickett / Papilio (tl); Martin Zwick / DanitaDelimont (tr); Neil Bowman (cra); Robert Pickett / Papilio (cl). **96 Alamy Stock Photo:** GM Photo Images (tc); Jos Mara Barres Manuel (l). **Dreamstime.com:** Alfio Scisetti (cb). **97 Alamy Stock Photo:** Stephan Bonneau / Biosphoto (tr); Zoonar / Olga Lipatova (cl); MJ Photography (b). **Dreamstime.com:** Shamils (tl). **98 Alamy Stock Photo:** imageBROKER.com GmbH & Co. KG / Therin-Weise (ca); Chris Lewington (cl). **98-99 Alamy Stock Photo:** Zdenk Mal. **99 Alamy Stock Photo:** Martin Zwick / DanitaDelimont (tr). **100-101 AWL Images:** Nigel Pavitt. **103 Dreamstime.com:** Xi Zhang (br). **Getty Images / iStock:** owngarden (cla). **104-105 Getty Images:** Guang Cao. **105 Dreamstime. com:** Eprom (ca). **106 Dreamstime.com:** EPhotocorp

(clb). **naturepl.com:** Gavin Maxwell (tl); Xi Zhinong (r).
107 Alamy Stock Photo: Joel Sartore Photography / Design Pics Inc (r); Wildscotphotos (tl); Ranjith Kumar (c). **108 Dreamstime.com:** Natallia Yaumenenka (c). **Shutterstock.com:** Captain Wang (bl). **108-109 Getty Images:** Yves ANDRE. **109 Dreamstime.com:** Natallia Yaumenenka (cr). **110 Alamy Stock Photo:** AY Images (bl). **Dreamstime.com:** Liz Lee (tr); Meinzahn (br). **111 Alamy Stock Photo:** Dong Lei / Nature Picture Library (bl). **John Birks:** (tl). **112-113 Shutterstock.com:** LinXiaoFu. **114-115 Dreamstime.com:** Leonid Andronov. **117 Shutterstock.com:** lavizzara (br). **119 Alamy Stock Photo:** Clement Philippe / Arterra Picture Library (cr); Boaz Rottem (b). **Dreamstime.com:** Ekaterina Kriminskaia (c). **Shutterstock.com:** Stephen Barnes (cl). **120 Alamy Stock Photo:** Christian Htter (bl). **Dreamstime.com:** Mikelane45 (tr); Anusorn Sutapan (br). **Shutterstock.com:** 23frogger (tl). **121 Alamy Stock Photo:** Hugh Harrop / AGAMI Photo Agency (tl); Rosanne Tackaberry (bl); Joseph Meng Huat Goh (r). **122 Alamy Stock Photo:** Bob Gibbons (bl); imageBROKER.com GmbH & Co. KG / Erhard Nerger (tr). **Shutterstock.com:** Tempus Aura Eugenie R (tl). **123 Alamy Stock Photo:** land-scaper / Panther Media GmbH (tl). **Shutterstock.com:** High Mountain (cl). **124-125 Alamy Stock Photo:** amana images inc.. **125 Alamy Stock Photo:** SHOSEI / Aflo Co., Ltd. (cr); Julie Pigula (tr). **Dreamstime.com:** Teerawat Winyarat (bc). **126-127 Getty Images:** Yuga Kurita. **129 Alamy Stock Photo:** Peter French / Design Pics,Pacific Stock (crb). **Dreamstime.com:** Manfred Thuerig (bl). **130-131 Getty Images:** Laszlo Podor (tc). **131 Alamy Stock Photo:** G. Brad Lewis / Photo Resource Hawaii (cra); C. Storz (b). **132 Alamy Stock Photo:** Fotofeeling / Westend61 GmbH (bl); David Ponton / Design Pics Inc (tl). **Alan Cressler:** (tr). **133 Alamy Stock Photo:** Michael Walther / Oahu Nature Tours Inc. (cl); Science History Images (tl); Jack Jeffrey / BIA / Minden Pictures (cr, bl). **Shutterstock.com:** Thomas Chlebecek (br). **134 Alamy Stock Photo:** Alvis Upitis / Pacific Stock / Design Pics Inc (bl); Michael Sznyi / imageBROKER.com GmbH & Co. KG (tl). **Dreamstime.com:** Christian Weber (crb). **134-135 Alamy Stock Photo:** Jack Jeffrey / Photo Resource Hawaii (tc). **135 Alamy Stock Photo:** Jack Jeffrey / Photo Resource Hawaii (tr, cl); Michael Sznyi / imageBROKER.com GmbH & Co. KG (br). **136 Alamy Stock Photo:** Jim West (c). **136-137 Shutterstock.com:** JKLovelacePhotography.

137 Shutterstock.com: Phillip B. Espinasse (cb); Sam Strickler (tr). **138-139 Alamy Stock Photo:** Alisa L. Gallant / USGS. **140-141 Alamy Stock Photo:** Last Refuge / robertharding. **143 Alamy Stock Photo:** Martin Harvey (bl). **Getty Images:** Image Source (tr). **144 Alamy Stock Photo:** blickwinkel / F. Neukirchen (b). **145 Alamy Stock Photo:** Flavio Varricchio / BrazilPhotos (b); Pulsar Imagens (tr). **146 Philippe J.R. Kok:** (tl, bc). **147 Alamy Stock Photo:** Ch'ien Lee / Minden Pictures (t). **148 Alamy Stock Photo:** Gabbro (tc); Ch'ien Lee / Minden Pictures (cl). **148-149 Alamy Stock Photo:** Nate Chappell / BIA / Minden Pictures. **149 Alamy Stock Photo:** Marc Guyt / AGAMI Photo Agency (bc). **150 Alamy Stock Photo:** anjahennern (br); blickwinkel / Neukirchen (cla); blickwinkel / F. Neukirchen (c). **Dreamstime.com:** Adwo (bl). **151 Alamy Stock Photo:** Westend61 GmbH (c). **Dreamstime.com:** Barbara Dymidziuk (bl); Piccaya (tr). **Shutterstock.com:** Vladimir Melnik (br). **152-153 Dreamstime.com:** Alicenerr

Cover images: *Front:* **Alamy Stock Photo:** Galaxiid bl, enricocacciafotografie / mauritius images GmbH b; **Getty Images:** Nisian Hughes t/ (Mountain); **Unsplash:** Art Lasovsky cb, Jerry Zhang t; *Back:* **Alamy Stock Photo:** enricocacciafotografie / mauritius images GmbH b; **Getty Images:** Nisian Hughes t/ (Mountain); **Unsplash:** Art Lasovsky cb, Jerry Zhang t

All other images © Dorling Kindersley Limited